The International Library of Bioethics

Volume 112

The *International Library of Bioethics* comprises international and interdisciplinary volumes on foundational and applied issues in bioethics. The series focuses on perennial topics of aging, mental health, preventive medicine, medical research issues, end of life, biolaw, and other areas of bioethics, whilst expanding into other current and future questions involving the living world. The *International Library of Bioethics* examines the role of bioethics against the background of increasing globalization and interdependency of the world's cultures and government.

The series explores meaningful ways to answer the questions of how to direct biotechnology to human and other living things' ends, how to deal with changed values in the areas of religion, society, and culture, and how to formulate a new way of thinking. That is, to create a new, practical bioethics.

The international Library of Bioethics welcomes book proposals representing the broad interest of the series' focus. We especially encourage proposals addressing aspects of changes in biological and medical research and clinical health care, health policy, medical and biotechnology, and other applied ethical areas involving living things, with an emphasis on those interventions and alterations that force us to re-examine foundational issues.

Brenda Bogaert

Affective Injustice in Healthcare

Springer

Brenda Bogaert
Institute of Humanities in Medicine
UNIL/CHUV
Lausanne, Vaud, Switzerland

ISSN 2662-9186 ISSN 2662-9194 (electronic)
The International Library of Bioethics
ISBN 978-3-031-94377-5 ISBN 978-3-031-94378-2 (eBook)
https://doi.org/10.1007/978-3-031-94378-2

This Springer imprint is published by the registered company Springer Nature Switzerland AG
The registered company address is: Gewerbestrasse 11, 6330 Cham, Switzerland

Acknowledgements I would like to thank my colleagues of the Healthcare Values Chair at the University of Lyon III (France) and at the Institute for Humanities in Medicine at the Lausanne University Hospital and the University of Lausanne (Switzerland) for many thoughtful discussions on the role of affects in healthcare throughout the years.

I would also like to thank the Brocher Foundation, which gave me the ideal working environment to finish this book and the opportunity to discuss the subject with an encouraging group of international and interdisciplinary colleagues working on issues of health and healthcare.

Competing Interests The author has no competing interests to declare that are relevant to the content of this manuscript.

Contents

Part I
Affective Injustice in Healthcare

Chapter 1
Introduction to Affective Injustice

Abstract In this introductory chapter, I will introduce the concept of affective injustice and how it will be used in this book. I define affective injustice in healthcare as those norms and practices that seek to minimize, hide, and silence affects. In this book, I will specifically be interested in how affective injustice happens to and affects healthcare providers, patients, and families. In this introductory chapter, therefore, it is necessary to first discuss how the concept is related to the concept of epistemic injustice and why the word injustice has been chosen to discuss this controversial subject. I will also introduce the key terms that will be used throughout this book, in particular affective related testimonial and hermeneutical injustice, affective harms, and affective silencing. I will also introduce terms that work toward affective injustice, including affective humility, affective resources and opportunities, affective freedoms, and affective recognition. This introductory chapter will prepare for the next two chapters to follow, which will discuss the affective harms experienced by healthcare providers and patients.

Keywords Affects · Affective injustice · Affective silencing · Affective humility · Epistemic injustice

1.1 Introduction

Healthcare is viewed as a practice based on rationality and objectivity. However, medicine involves working with human beings…by other human beings. It is practice inseparable from relationships between the patient, their family and the healthcare team (Heyhoe et al 2016). Because of this, healthcare professionals' work is intertwined with affects. Patients and families experience joy when they have cancer remission announced to them by their doctor, but the doctor too feels joy for them and for themselves as professionals when they are able to help them. In the same way that patients feel fear and anxiety in the face of the unknown, these affects also haunt healthcare workers in the face of uncertainty when they do not know what to do to help their patient, or when they have done all that they can and it is still not

B. Bogaert, *Affective Injustice in Healthcare*, The International Library of Bioethics 112, https://doi.org/10.1007/978-3-031-94378-2_1

enough. Shame and disgust too are integral parts of healthcare work, affecting how healthcare providers treat patients and their capability to cope (Jaeb and Pecanac 2024, Hadjittofi et al 2022).

Because they experience many emotionally challenging situations in their professional lives, it is important for healthcare providers to find ways to accept, express, and deal with affects in order to foster their well-being and resilience and to prevent burnout (Weilenmann et al 2018). Given the high rates of burnout, depression, substance abuse and suicide compared to the general population (Ryan et al 2023; Shah et al 2021), this is an urgent public health problem, one that affects not only healthcare providers' well-being, but also clinical outcomes, including quality of care, patient satisfaction, and treatment adherence.

Even though affects are an unavoidable part of healthcare work, they are still viewed with suspicion and apprehension. There is a persistent idea—held by healthcare providers, educators, and institutions—that healthcare providers should move beyond their affects to become better professionals. As we will see in this book, healthcare providers have been encouraged to "hide" or "minimize" their affects in order to encourage effective communication and relationships. Yet clinical practice, which involves reflection and deliberation to make decisions, is not merely based upon cold thinking alone. Decisions are based upon how healthcare providers feel about their patients and how they interact with them. In the same way, in interdisciplinary meetings, while there is a strain on team members "not to become emotional" to permit collective reflection, affects about their patients are not only an important means for them to come together, but can also help alert to institutional dysfunction. In the same way, in training, future doctors are taught to keep their emotions out of healthcare work to stay professional in front of faculty, their peers, and future patients. Yet these interpersonal relationships, how they feel about their patients and their desire to help them, are the very reason that doctors and other healthcare professionals choose their profession in the first place. Why then do affects continue to be viewed as "irrational" or "unprofessional"? Even though medical institutions and healthcare providers have tried to put aside the emotional work of healthcare, it forms an important means of making them caring, capable healthcare providers. As I will argue in this book, affects may—in many circumstances, although of course not all—help healthcare providers to make better decisions. It may also motivate them to work harder for their patients, as it helps them focus on what is important to them and the care relationship. This is because affects are a reflection of their values, of what is important to them and how they feel about themselves as professionals.

In return, while it may be thought that the patient's emotions have been more readily accepted in the turn toward person-centered care and more empathy-based medicine, as we will see in this book, this is not necessarily (or at least not always) the case. The idea that patients may be "too emotional" to decide important healthcare decisions leads to pressures on them to suppress their affects in order to be taken seriously. Furthermore, the idea that some patients are too emotional (or that their emotions are somehow exaggerated or not adapted to the situation at hand), in particular affects certain groups, particularly women and disadvantaged groups, who may already face difficulties in having their illnesses and care experiences be heard and understood.

This book argues for an affective turn in healthcare. As I will show, by not giving affects their proper place, we have largely sidelined and silenced an essential part of healthcare practice. We have also facilitated the burn-out of the medical profession as well as encouraged patient dissatisfaction and isolation. I will argue that it is necessary to find safe spaces for the expression of affects, be it for the patient, their families, or their healthcare providers, and I will do so by linking the subject with the emerging concept of affective injustice. To make this subject not (only) a description of the problem, in Part II of this book, I will also look at possible remedial strategies. However, to prepare the discussions that will follow, I will first elaborate the concept of affective injustice to better understand its articulation in the context of this debate.

1.2 Introducing the Concept of Affective Injustice

In our journey, I will use the concept of affective injustice to elaborate the subject of the ambiguous and polemic place for affects in healthcare. The term affective injustice has started to gain significant interest in the philosophical academic community since at least 2018, with influential publications by philosophers Amia Srinivasan (2018) and Shiloh Whitney (2018). The term has since achieved more conceptual precision and a close rapprochement to epistemic injustice via the works of Archer and Mills (2019) and Gallegos (2021). However, it is to be acknowledged that its discussions and themes can be traced to ongoing discussions that have ebbed and flowed on the place for emotions in social life in the social and human sciences for decades, in particular thanks to the so-called "emotional turn" in these fields, which has given greater weight and attention to the role of affects.

As the affective injustice debate expands, its contours, possibilities, and limits for understanding the role of affects in our social lives will be clarified and developed. For now, authors have focused on social movements and in particular how silencing affects has become another means of excluding marginalized groups. While the affective injustice debate is expanding as I write these words, in this book, I will primarily be interested in what affective harms may occur for healthcare providers, patients, and families in healthcare and what we can do to rectify these problems. This book therefore aims to give conceptual precision and theoretical concepts to this expanding debate; however, it is also concerned with the subject in a more practical way, namely finding ways of working toward greater affective justice for healthcare actors.

In the debate on affective injustice, I have been notably inspired by the characterizations advanced by Srinivasan (2018), who first conceptualized the idea of affective injustice, that of Archer and Mills (2019), who furthered Srinivasan's characterization by linking it with the concept of emotional regulation, and Gallegos (2021) who conceptualizes what affective justice may look like, notably through guarantees of affective opportunities, resources, freedoms and recognition. In this section, I will briefly introduce these concepts to help prepare the discussion to follow.

Table 1.1 summarizes the most relevant concepts and terms that will be used in this book. It is to be noted that the concept of affective injustice is developing quickly and there are a number of other terms currently in use. The table below reflects those most relevant to the discussion of affective injustice in healthcare, including the concepts that I have introduced in this book. It is to be noted that the table below is not meant to be exhaustive of all possible concepts/terms.

Amia Srinivasan (2018) in the first conceptualization of affective injustice, helped clarify the reasons why some persons may be affectively silenced. This is, according to her, notably because others (including persons from their own social group) see these affects as inappropriate for one reason or another. She gives the example of Black activists in the United States, who in the face of racism and police violence, are often told not to angry—including by influential Black leaders themselves—because showing their anger is seen as jeopardizing the dialogue that they are trying to achieve. This situation leads to what Srinivasan calls a "psychic tax" on the person oppressed, as they are forced to choose between the legitimate expression of their feelings and the ways it is considered socially acceptable for them to act. As the philosopher shows, the affective harms from not recognizing or acknowledging such emotions not only negatively influence the person who is silenced, but it also has wider effects on oppressed groups fighting for change. She argues anger in particular can be a powerful motivating force and rally persons together, as well as clarify what issues are at stake. Srinivasan therefore shows both the individual and collective harms that come from silencing affects.

Before going further, to clarify, my book is not focused on social movements per say, although the affective injustice dimensions of collective action in healthcare merit further analysis, as I have argued in other work (Bogaert 2022). However, what Srinivasan brings to the table is a multifactorial analysis on both the individual and social dimensions of the problem, and in particular of the specific harms that may come from silencing affects. In the context of healthcare, I will investigate the individual harms that may come to patients or healthcare providers, but also how it may negatively affect healthcare outcomes, patient satisfaction, and healthcare provider-patient relationships.

However, Srinivasan's conceptualization of affective injustice is inspiring for another reason. In particular, it turns the table on affects that are classically viewed as negative—in her example, anger—to show their value to help the person and the group to focus on what is important. In my discussion of affective injustice in healthcare, I will frequently take this approach when discussing so-called negative emotions (fear, anger) that are often devalued and seen as counterproductive to care. Like Srinivasan, I show that these negative emotions may help crystalize what is of value to the person, and as I will demonstrate, say something important that may be missing from the case file or alert to dysfunctions in the team. I will show for instance how the patient's anger about their illness is not necessarily a bad thing, as it is related to changes in relationships and life stories that needs to be taken seriously to better accompany them in their illness journey. I also show that we should not necessarily

Table 1.1 Concepts of affective injustice and related terms

Concept	Definition
Affective injustice	Definitions currently include: – An injustice faced by someone specifically in their capacity as an affective being (Archer and Mills 2019) – A state in which individuals or groups are deprived of affective goods which are owed to them (Gallegos 2021) – When treatment of emotions is unjust, or emotions are used to treat people unjustly (Pismenny et al 2024)
Affective injustice in healthcare	Norms and practices that seek to minimize, hide, or silence affects in healthcare providers, patients, or families
Related and complementary terms	
Affective neutrality	The expectation that the physician strives for and maintains objectivity in relation to patients by not expressing affects (Parsons 1951)
Affect-related testimonial injustice	A person or group is unfairly subjected to credibility deficits, silencing, smothering—that is, the creation of conditions that strongly encourage or require self-silencing and other practices and circumstances that undermine how others give uptake to their affective assessments of things and their views about which emotional responses may be warranted (Gallegos 2021)
Affect-related hermeneutical injustice	A person or group's ability to interpret their own and others' affective experiences and emotional responses is unfairly constrained or undermined, for example, by gaslighting or by the unavailability of concepts that would allow them to be understood(Gallegos 2021)
Emotional disorientation	When persons develop doubt about what they value and experience due to downgrading, ignoring, or misunderstanding their affects (Gallegos 2021)
Emotional pathologizing	Emotions/affects are distorted by being viewed through a medical lens, as they are regarded as symptoms of a psychiatric disturbance (Pismenny et al 2024)
Affective harms	Harms related to well-being (Srinivasan 2018)
Affective silencing	Attempts to silence others' affects when we do not wish to hear or understand them due to a prejudice or social norms/expectations
Affective self-silencing	Deliberate attempts to silence one's affects due to a belief that one's affects will be not heard or understood, or due to affective norms/ expectations
What we are seeking	
Affective justice	The capability to exercise affective resources, opportunities, freedoms, and forms of recognition (Gallegos 2021)
Affective freedoms	Freedom from interference in the pursuit of subjective well-being, including freedom from circumstances that give rise to emotional distress and negative or unpleasant emotions and moods (Gallegos 2021)

(continued)

Table 1.1 (continued)

Concept	Definition
Affective resources and opportunities	Materials, activities, and circumstances that contribute positively to one's subjective well-being, including nurturing interpersonal and social relationships; sleep, therapy, and other means of providing self-care; and affective scaffolds in the built environment that facilitate positive mood and self-evaluation (Gallegos 2021)
Affective recognition	Respectful consideration of, and responsiveness to, one's particular needs with regard to subjective well-being (Gallegos 2021)
Affective humility	The awareness that we do not necessarily experience and express the world in the same ways and an active disposition to look for and understand another's affective experiences
Affective capabilities	The capability (the resources and opportunities) for the person to be able to develop, show, and have their affective experiences and assessments understood and recognized by others

ignore or put aside a healthcare provider's anger, for instance in an interdisciplinary meeting, as it may alert to potential burn-out or team dysfunction.

The second work important for this characterization of affective injustice comes from Archer and Mills (2019). In their article, they both link the subject to epistemic injustice and to emotional regulation studies in psychology. In the first instance, they show how affective injustice is closely related to epistemic injustice (a wrong done to someone in their capacity as a knower) by arguing that affects are a form of knowledge. This is an important consideration for healthcare, as it means that if we ignore affects, we may be missing important information on patients (that may be necessary for diagnosis and care) and other facets of healthcare provider experiences that can help improve institutions and interprofessional coordination. I will come back to this point regularly in the book, as it is important to understand the extent of the individual and collective epistemic harms that come from silencing affects.

Archer and Mills however specify that the concept of affective injustice must not be seen merely as a side topic of epistemic injustice, as it may involve specific harms that go beyond (or are separate to) epistemic injustice. As they argue, the affective harms that may come out of affective injustice are related specifically to well-being, which is a wider topic than (just) epistemic justice, although it is a frequent outcome. This point will be important to Part II of this book, when I look at the means to achieve affective injustice (notably through affective resources, opportunities, freedoms, and recognition). As I will show, this may include the creation of safe spaces that enable knowledge sharing and receiving, but it will also include initiatives that foster greater well-being.

Finally, Archer and Mills' article also makes an important theoretical innovation by linking affective injustice to studies on emotional regulation. Emotional regulation strategies, which can be broadly described as the ways people manage their emotional lives, is an important aspect of how affective injustice happens in practice, and in particular in healthcare, as I will show. While a considerable amount of research has shown how such kinds of emotional regulation strategies may be useful for healthcare providers, I will also show that they can also lead to a number of harms for the person (in addition to doing little to change the situation). Although I will explore various strategies and their effects in this book, I will give here an example to better understand this debate. Healthcare providers experience a variety of affects in their healthcare practices: fear of making a mistake; of not being able to help their patients; anger toward patients when they are aggressive with them or toward the institution when they are unable to provide quality care; and sadness at the death of their patient. At certain times, emotional regulation strategies may be necessary to help them remain calm in the face of challenging situations; however, as I will discuss, they often do little to (and may even hurt) the provision of quality care and healthcare provider well-being. By promoting hiding behaviors or distraction strategies, they are likely to negatively influence the healthcare provider (who needs to find an outlet and a safe space for discussion), as well as discouraging effective communication with others. I will discuss these issues in greater depth in the next chapter, but suffice to say here that the link between affective injustice and emotional regulation will be important to understand the extent of affective harms experienced by healthcare providers, patients, and families.

The final author to provide significant theoretical support to this project is Gallegos (2021), notably by helping to better understand what we are seeking in the affective injustice debate as well as introducing the terms affect-related testimonial and hermeneutical injustice. First of all, he defines affect-related testimonial injustice as those situations in which a person or group is subject to credibility deficits, silencing, and smothering in their affective assessments. In a similar way, he defines affect-related hermeneutical injustice as, "a person or group's ability to interpret their own and others' affective experiences and emotional responses is unfairly constrained or undermined, for example, by gaslighting or by the unavailability of concepts that would allow them to understand (Gallegos 2021)." Similar to the epistemic injustice discussion, therefore, we can see the individual and collective harms of both types of affective injustice. They stem from prejudice and bias; they come from a lacune in the collective understanding; and they tend to affect non-dominant groups. They also lead to specific harms, from emotional disorientation to confusion and isolation for both the person and the community. It also means missing out on valuable information and experiences that can benefit everyone.

Gallegos however does not stop there. He defines a number of goods that should be ensured to achieve affective justice. These are freedoms, resources, opportunities and forms of recognition (2021, p. 189). He therefore helps us better understand the end goal and what we are seeking in this debate. It is the following:

- *Affective freedoms*: freedom from interference in the pursuit of subjective well-being, including from circumstances that give rise to emotional distress and negative emotions and moods

- *Affective resources and opportunities*: materials, activities, and circumstances that contribute positively to one's subjective well-being, including nurturing interpersonal and social relationships; sleep, therapy, and other means of providing self-care; and affective scaffolds in the built environment that facilitate positive mood and self-evaluation
- *Affective recognition*: respectful consideration of, and responsiveness to, one's particular needs (and a recognition that one's affects are legitimate)

As I will explore in this book, healthcare providers are currently missing many of these resources, opportunities, freedoms, and forms of recognition. In terms of affective resources and opportunities, although self-care and safe spaces may be increasing in some healthcare institutions, having the space to express one's affects has largely been constrained to young professionals just coming out of medical school (in particular via the so-called analysis of practice) and are almost exclusively geared for young doctors. In addition to this focus on age, which assumes that more experienced doctors do not need emotional support, nurses and other paramedical professionals also suffer from affective injustice and need such safe spaces.

In terms of affective freedoms, healthcare providers—in particular nurses—are often subject to moral distress, as they feel helpless to change the situation as they are often not in charge of care decisions. This is not only an individual or interpersonal problem but concerns a wider problem of hospital hierarchies and change that will need to be addressed not only through safe spaces, but also greater recognition of the unique contributions of these professionals toward patient care.

Finally, it will also be important to support greater affective recognition. Paying attention to one's needs—and one's affects–means taking them seriously and not dismissing them as "irritational." Giving the necessary recognition to affects also means responding to healthcare providers' needs and what these affects may say about their relationships with their patients and their institutions, and in turn of patients about their care and illness journeys. In the meantime, being silenced or sanctioned when trying to express one's affects still remains the standard in healthcare institutions today.

These are the main concepts that have inspired my understanding of and conceptualization of affective injustice in healthcare, although I will also refer to other emerging concepts in the evolving affective injustice literature (as shown in Table 1.1). In the meantime, there are also a number of concepts from the epistemic injustice literature that will also be helpful to clarify and deepen the discussion. One of these is silencing (Dotson 2011; Fricker 2009). In her original conception of epistemic injustice, Miranda Fricker shows that are a variety of ways in which persons are silenced. There are deliberate attempts to silence others who we do not wish to hear or understood due to a prejudice. However, there are also other situations in which the person will evaluate a situation in advance and if they believe that they will be misunderstood or misheard, they may decide to silence themselves. Dotson (2011) calls this problem testimonial smothering. As I will show, in the debate on affective injustice in healthcare, while there are rarely examples of deliberate attempts to silence other's affects (although this happens), self-silencing/smothering of affects frequently happens as healthcare providers, patients, or families understand that they will be judged unfairly. As the epistemic debate has shown, silencing and self-silencing are particularly harmful to the person, and they often isolate them from others, put doubt on their experiences, and may negatively influence their well-being. Therefore, *affective silencing* will be an important part of my account of affective injustice's articulation in healthcare.

Another related concept that I will adapt from the epistemic injustice literature is epistemic humility (Wardrope 2015). I define *affective humility* as an individual engagement of the person to understand the limits of their affective knowledge and an active disposition to seek out another's experiences. The concept of epistemic humility has its origins as part of the discussion of epistemic injustice—but it is also one that is discussed in a wider context of the need for greater humility in medicine in order to better recognize patient experiential knowledge along with (not in hierarchy) the healthcare provider's expertise. The concept of affective humility will be developed to better understand how healthcare providers may develop and cultivate it in their practice, thereby providing a safer affective space for others. A call for affective humility is also a rallying call to give greater voice and recognition of the other, including their affects. This concept is closely related to Halpern's concept of curiosity (2001, p. 129), which focuses on suspending judgement and being attentive to what patients communicate both verbally and non-verbally and of recognizing the limits of our understanding of the other's emotional states. I will come back to this issue in Part II, as Halpern's proposal notably concerns the time needed for patient narratives, as well as the necessary modesty in the face of different ways of experiencing and viewing the world.

Another term that will be used has been inspired by the capabilities approach and in particular Martha Nussbaum's approach (Nussbaum 2013). The term is affective capability, adapted here from Martha Nussbaum's list of central capabilities. By this capability, I speak of the capability (the resources and opportunities) for the person to be able to develop, show, and have their affective experiences and assessments understood and recognized by others. By using the term capability, I am voluntarily putting the focus on the person's capacity to act but also on the social environment in which they act, which is an important consideration in how affective injustice actually happens in practice. Having and being able to exercise affective capabilities is a key ingredient of ensuring affective justice for healthcare actors, as we will see. It involves developing capabilities, in order to effectively deal with difficult care situations, but also the ability to exercise these capabilities in interactions with healthcare institutions and with their colleagues.[1]

As I will show, there are currently intentional and unintentional restraints on affective expression in healthcare, which are complex and difficult to unravel, as they are based on recurring ideas of healthcare providers as "objective" and "non vulnerable." There are also pressures on patients to suppress affects when discussing

---

[1] Stockdale (2024) in a similar way has claimed that affective injustice occurs when the person's abilities to develop and exercise human capacities and skills has been hindered, although she does not make explicit reference to the capabilities approach.

healthcare plans in order to make "rational" decisions. What I am seeking by using the term affective capability is both the capability to find resources and opportunities to express affects but also a personal acceptation that affects are important to help understand and articulate care experiences. This is an important point given the pressures that healthcare providers have internalized in their training, which make them believe that affects are somehow unprofessional (see Part II). In addition, by using the world capabilities, I am also making a deliberate attempt to put the approach under the capabilities approach's social justice framework, which is intended to facilitate the abilities of all persons to be able to develop and exercise their capabilities. Furthermore, to use the capabilities approach language, by recognizing affects as a core capability, I understand the full extent to which affective expression may be important for the person to flourish.

Having now introduced the main concepts that will be important to the debate on affective injustice in this book, I will now return briefly to the long debate on whether or not emotions are cognitive, in order to clarify how I will see and use them in this book.

1.3 Are Emotions Cognitive?

In this book, I will favor the word affect over emotion, as it is the term that encompasses a range of feelings that the person experiences, as well as makes a clear link to the existing affective injustice literature. While other terms are emerging, such as emotional injustice (Pismenny et al 2024), the benefit of the more global term is that it includes both emotions and moods and can thus describe the wide variety of experiences that healthcare providers, patients, and families experience. However, the question then becomes, to what extent are they connected to cognition and what does this imply for our discussion?

As we will see, one of the factors influencing the negative reception of affects in healthcare is the conception of them as irrational. In particular, there is a recurring idea that affects will interfere with medical judgement, which is supposed to be objective/neutral and based upon "cold" rational thinking alone. There is the idea that particularly strong emotions are disruptive to sound decision making, such as when a clinician who becomes angry has trouble concentrating. However, there is also the perception that even less strong feelings such as frustration or anxiety are unreliable sources of information because they remain subjective experiences (Halpern 2001, p. 30). Both ideas are based upon a long philosophical tradition that holds that reason and affects are opposites. While philosophical views have significantly changed in recent years and most philosophers now agree that emotions involve some cognitive aspects, this view has not (yet) been incorporated into most medical thinking.

If we take this further, following Martha Nussbaum (2003), emotions are part of our ethical reasoning. She says that emotions are appraisals or value judgements of the object toward which it is directed. This means that the object of the emotion may show us something important which is needed for the person's flourishing. In this book, I will advocate that healthcare provider and patient affects are important sources and forms of knowledge. They show something of salience to the person—about their care, about their lives with illness, or their well-being in their professions—and thus cannot be ignored. Nussbaum shows for example that we do not show strong emotions like fear over things or persons that are unimportant to us: the very intensity of the emotion shows that something of value is at stake. While we can see this emotion may be directed toward some immediate danger important to our survival (such as being hit by a car), it may also be related to something that we are afraid of losing that is of particular value to us (such as our family member). Taken in this way, even negative emotions like fear or anger are important and can provide important information about aspects of care. Therefore, what I am interested in are those affects, following Gallegos (2021), that show something of particular importance to the person.

In this book, I will therefore be most interested in those situations where the healthcare actor is somehow prevented from expressing their affect or having it recognized due to social norms and expectations about what is appropriate behavior or based upon a belief that they are irrational or not important. In the context of healthcare, these can be internal social norms that have been learned in society, and specifically in medical training and practice, but they can also stem from external pressures by colleagues and healthcare institutions to regulate or even suppress them, as they are viewed as "unprofessional." As we will see, these affects often show something of importance about the care relationship or their interprofessional roles that merits being taken seriously. Therefore, in this book, I am primarily interested in exploring those situations in which external pressures prevent it from having uptake (Whitney 2018, p. 495), or in other words when it is not appropriately moving to others. The perceived idea that it will not have uptake may also prevent the affect from being expressed in the first place, as our examples of affective silencing will show.

However, a word of caution is necessary. Gallegos has claimed that even if not all affective suppression is necessarily harmful, we should not be overly demanding in deciding that only the most severe forms of emotional abuse and deprivation should be deemed unjust (Gallegos 2021, p. 191). As I will argue in this book, healthcare providers and patients can be subject to many small, implicit, and even self-inflicted forms of affective injustice, many of which harm them as affective beings in small (and thus difficult to address) ways. From a public health perspective, these micro instances of affective injustice are not to be taken lightly, notably because they affect care quality and influence patient satisfaction and may be a factor in healthcare provider burn-out. Therefore, I will take the stance that most healthcare provider, patient, or family member's affects are to be taken seriously, and that we should pay particular attention if they show something about how their view their illness, their relationship with a patient, or their well-being.

1.4 Individual and Collective Dimensions of Affective Injustice

It is now necessary to discuss the individual and collective dimensions of affective injustice. In this book, I advocate the need to take a global view to understand its depth and complexity in healthcare. In the affective injustice literature thus far, researchers have focused on the different ways that affective harms may occur, from psychological perspectives (Srinivasan 2018), to social (Archer and Mills 2019), and embodied experiences (Whitney 2018). For instance, while psychological accounts may focus on the affective harm to the individual, social accounts center on explicit/implicit emotional regulation strategies that suppress them, and finally embodied accounts on how persons experience them in their bodies. As Krueger has shown (2024), while these perspectives may lead to different forms of analysis, they often intersect and are not necessarily in opposition. They are different aspects of a complex, multidimensional phenomenon.

The benefit of taking this global approach is to promote a more global and complex view of how affects are received and understood by healthcare actors, but also to incorporate the influence of the environment. This will also help to achieve its theoretical innovation and to enable it to incorporate other perspectives. There already exists already an extensive intellectual canon on affects and how they are received and understood. Scholars such as Stockdale (2024) have sought to distinguish affective injustice from terms such as psychological oppression by focusing on its more structural dimensions; others (Archer and Mills 2019) have sought to benefit from studies in emotional regulation to better understand social dynamics. I take the stance that it is important to acknowledge and build upon the psychological and sociological heritages coherent with the affective justice framework but also attempt to go further to understand the intersection and complex nature of this phenomenon.

Therefore, I will investigate the individual affective harms that may come from individual emotional suppression strategies. However, I will also seek to understand in particular how certain groups—such as doctors—are emotionally socialized to suppress their affects because of social norms and expectations. In Part II of this book, I will also take a more ecological perspective and determine how hospital design and architecture may influence affective injustice, a perspective which is complementary to Kruger's analysis (2024) of the influence of hostile architecture on well-being and capabilities. These diverse levels of analysis will also lead me to consider how affective practices may affect care quality and care outcomes. Taking together the individual, social/collective, and embodied/ecological dimensions of this topic therefore give it a richness and a complexity, as well as help understand its salience to healthcare.

1.5 The Use of the Word "Injustice"

I have now described the key terms and ideas used in this book to explore affective injustice. However, the question now becomes whether or not the word injustice is appropriate to describe these situations. As discussed, the term affective injustice has thus far been primarily applied to social justice movements, where the link with the term "justice" seems more obvious, as it refers to movements to overcome oppression against marginalized groups. Stockdale (2024) has proposed that the demand for a person to regulate their emotions is a matter of justice insofar as it is an "affective practice" that participates in social conditions of oppression. It may therefore seem surprising to readers that I am proposing to use the concept in the context of healthcare.

First of all, I should clarify that I do not see healthcare providers or patients as necessarily belonging to an oppressed group in a wider sense, or at least not in the same way as those facing racism and police violence. However, as the extensive literature on epistemic injustice has shown, persons may become subject to a deflated level of credibility depending on their social status in certain situations. The epistemic injustice discussion has shown for example that the status of a patient (in some circumstances) may be a criterion for prejudice such as when patients are considered too "emotional" for their testimony to be considered valuable (Carel and Kidd 2014), while other researchers have highlighted that hospital hierarchies may also function in a similar way, such as when a nurse is taken less seriously than a doctor in an interdisciplinary meeting (Reed and Rishal 2015). These persons are not necessarily considered oppressed in their wider social environment; however, they are oppressed owing to their social statuses in these specific situations. As I will explore in this book, patients and healthcare providers may be subject to affective injustice due to their social statuses and the specific expectations that we have about them, in particular expectations that they remain objective (unemotional). Think of the nurse who is expected to put on their "professional face" and be positive and cheerful even in front of an aggressive patient, or the doctor who is expected to remain affectively neutral to excel in their job. In making this rapprochement, I am not making light of the experiences of oppression that marginalized groups live with everyday, nor the serious affective and epistemic harms that impact their flourishing; however, some persons—even with a high social status and who are considered epistemic authorities (namely doctors)—may be harmed when they cannot make sense of their affective experiences, or when their affects are downgraded, discouraged, or not taken seriously. As Stockdale (2024) argues, affective norms, practices, and relationships may affect us all because we have structured society in such a way that many aspects of living in society benefit no one. This means that affective injustice can also occur for those who are often seen as privileged.

If we can accept that the social status of the healthcare provider or the patient can be a factor of prejudice, a question still however remains whether or not we should use the word "justice" to talk about this issue. Since the influential publication of John's Rawls' Theory of Justice (1997), the word "injustice" has become a popular term in both moral and political philosophy and has been used widely, from healthcare to ecology discussions and beyond. However, its use is coming under increased scrutiny and there is a question if it is being misused. Plunkett (2021) notably claims that the way philosophers talk about "justice" makes it hard to advance normative arguments in determining how we should set up and run our social and political institutions.

I defend this choice for two reasons. First of all, there is a recurrent problem in the reception and acceptation of the word "justice" into urgent social justice discussions. Those who are against using the word are (also) those not necessarily the ones who are fighting for change (and may even be satisfied with how things are). Using the word "justice" therefore is strategic: it is a call for change. It brings attention to the situation and the seriousness of the situation. It is also necessarily provocative. Given the current crisis facing healthcare institutions, a crisis "bigger than the COVID crisis," due to staff quitting and closure of services, taking care of affects is not only an academic problem, but also an urgent public health problem. Using the word injustice in this case therefore seems appropriate. The second point is in reference to Plunkett's criticism that using the term injustice may not help guide decisions. As I will show, the affective injustice debate can help us to find a number of remedial strategies that give clear guidance. Therefore, using the term injustice will not only help show the urgency of the problem, but will also allow me to take a modestly normative approach (Spranzi 2018) to propose solutions to these problems.

Another reason that I have favored the word injustice in this discussion is to make a clear link to the wider epistemic injustice discussion. As I will explore in this book, many (although not all) situations of affective injustice have repercussions in how the actor's knowledge is recognized and understood. As I have argued, affects are forms of knowledge and therefore not to be discounted. For instance, when a doctor does not take seriously a patient's testimony as it is viewed as too emotional, this means that they may lose out on valuable information that may help them to make a diagnosis or to decide with the patient the treatment plan. Similarly, when a nurse is not taken seriously in an interdisciplinary meeting as they were "too emotional," we may be missing out on information that could help guide interprofessional decision-making.

However, given that this account of affective injustice is specifically related to healthcare, it will take a narrower approach than most definitions of affective injustice thus far. By affective injustice, I am interested in those **norms and practices that seek to minimize, hide, or silence affects in healthcare providers, patients, or families.** Just as in the epistemic injustice debate in healthcare, I am both interested in the macro level (systematic factors) of exclusion but also those micro interactions such as between the healthcare provider and the patient in the consultation.

1.6 How the Book Will Elaborate Affective Injustice in Healthcare

Having described the key terms and concepts that I will use in this book, I will now elaborate the process of exploring affective injustice in this book. In a first part, I will focus on the various ways in which healthcare providers have been subject to affective injustice and the harms it causes them. I will in particular discuss how affects came to be viewed with suspicion in the medical field and the various ways that healthcare providers have sought to deal with this, notably by self-silencing their affects and encouraging silencing in others.

I will then focus on patients and families and the ways in which they are subject to affective injustice. This subject has been partly conceptualized in the epistemic injustice literature, notably by showing instances in which the patient's testimony is ignored as they are viewed as "too emotional," however, the subject merits further exploration in light of developments in the affective injustice approach. The focus of this chapter will be to better understand the ways in which affective injustice occurs for patients and their families and leads in some cases to substandard care. I will also start to explore some remedial strategies (and their limitations) in this chapter, particularly those proposed from the epistemic injustice literature. In this exploration, I will also not forget support persons and the specific affective harms they may experience in their loved ones' healthcare journey.

The goal of these two chapters is to set the stage, to show the problem and the extent of the problem. It will show that affective injustice is recurrent, in various ways and forms, for all healthcare actors, and that it affects care quality, well-being, and healthcare-provider patient relationships.

Having described the problem, the second part of the book will take a more proactive stance by looking at the various ways and means we can work toward greater affective justice. These chapters will be inspired by Gallegos' conceptualizations of affective resources, opportunities, freedoms, and recognition. They will be dedicated to various promising (if marginal) methods to give a greater space for affects into medical training and practice.

I will first investigate the promising place for narrative methods to give a safe space for affects. This chapter will primarily concern how narratives may permit a greater affective space for healthcare professionals. I will discuss several initiatives both from my own research and from other inspiring initiatives, notably the use of free-form journals, narrative medicine workshops, narrative banks, and the narratives created through interdisciplinary reflection. I will explore the possibilities and limits of each one to create a safe affective space for healthcare providers and their reception and use in institutions.

In the next chapter I will look at the place for spiritual care as an affective resource. The aim will be to show how an attention to and methodology for spiritual care may provide a safe space. My focus in this chapter will principally be on how spiritual care may provide safer affective spaces for patients, although opportunities for healthcare providers will also be explored.

Moving on from this debate, I will then show the possibilities and limits of patient expert/partner integration in hospital cadres, and the ways in which their inclusion may facilitate affective resources and opportunities for patients, in particular as a means of having their knowledge be better heard and understood. In order to nuance this debate, I will also discuss the potential harms that may come from this form of peer support in order to plan ways in which they may contribute positively.

I will then take a sidestep and investigate the importance of hospital design and architecture in creating affective safe spaces for hospital users. While the discussion will principally be focused on the patients' and families' experiences of the hospital space, in particular in the hospital room and waiting rooms, the affective needs of healthcare providers will not be ignored. Through a discussion on the various spaces of the hospital, as well as the influence of light, color, and sound, this chapter will help to see and understand the importance of design and architecture on the user's affective experience. However, it is to be signaled that this chapter is meant as a complement to the other methods proposed here: in other words, even with attention to hospital space and design, without a care relationship that is supportive and able to recognize the person's affects, it is unlikely the patient will experience this space as safe, despite the innovations made possible by hospital designers and architects.

Next, I will turn to the problems of affective injustice in medical training and discuss several methods that give a greater space for affects. I will contextualize this debate by starting from classic works in medical sociology and work up to the present day to show the (still) reluctant place that affects have in medical training. I will then discuss several promising methods, namely analysis of practice and role playing, which are already present in medical curriculum but may be adapted to provide a greater space for affects. A final part of the chapter will return to the question of affective humility and show how to work toward this virtue in medical students.

Finally, in the conclusion I will discuss the need for empirical research to move the debate on affective injustice forward. I will also give some ideas of possible research methods and areas of focus. This book is meant as a first step, a means to move toward greater awareness of the harms that come from ignoring or sidelining affects in healthcare. However, it does not just seek to highlight the problem, but also encourages us all to think creatively about what we can do about it, whether as philosophers, social scientists, educators, doctors, nurses, or concerned citizens.

References

Archer, Alfred, and Georgina Mills. 2019. "Anger, Affective Injustice, and Emotion Regulation." *Philosophical Topics* 47 (2): 75–94.

Bogaert, Brenda. 2022. "La colère des soignants : vers une justice affective dans l'espace public." *Savoirs en lien*, no. 1 (December). https://doi.org/10.58335/sel.173.

Carel, Havi, and Ian James Kidd. 2014. Epistemic injustice in healthcare: A philosophical analysis. *Medicine, Health Care and Philosophy* 17: 529–540. https://doi.org/10.1007/s11019-014-9560-2.

Dotson, Kristie. 2011. "Tracking Epistemic Violence, Tracking Practices of Silencing." *Hypatia* 26 (2): 236–57.
Fricker, Miranda. 2009. *Epistemic Injustice: Power and the Ethics of Knowing*. 1. publ. in paperback. Oxford: Oxford University Press.
Gallegos, Francisco. 2021. "Affective Injustice and Fundamental Affective Goods." *Journal of Social Philosophy* n/a (n/a). https://doi.org/10.1111/josp.12428.
Hadjittofi, Marilena, Kate Gleeson, and Anne Arber. 2022. "The Experience of Disgust by Nursing and Midwifery Students: An Interpretative Phenomenological Approach Study." *Nursing Inquiry* 29 (2): e12427. https://doi.org/10.1111/nin.12427.
Halpern, Jodi. 2001. *From Detached Concern to Empathy: Humanizing Medical Practice*. Oxford: Oxford University Press.
Heyhoe, Jane, Yvonne Birks, Reema Harrison, Jane K O'Hara, Alison Cracknell, and Rebecca Lawton. 2016. "The Role of Emotion in Patient Safety: Are We Brave Enough to Scratch beneath the Surface?" *Journal of the Royal Society of Medicine* 109 (2): 52–58. https://doi.org/10.1177/0141076815620614.
Jaeb, Michael A., and Kristen E. Pecanac. 2024. "Shame in Patient-Health Professional Encounters: A Scoping Review." *International Journal of Mental Health Nursing* (March), inm.13323. https://doi.org/10.1111/inm.13323.
Krueger, Joel. 2024. "An Ecological Approach to Affective Injustice." *Philosophical Topics*.
Nussbaum, Martha C. 2003. *Upheavals of Thought: The Intelligence of Emotions*. 1. paperback ed. Cambridge: Cambridge Univ. Press.
Martha Craven Nussbaum 2013 Creating Capabilities: The Human Development Approach. First Harvard University Press paperback The Belknap Press of Harvard University Press Cambridge, Massachusetts London, England
Talcott Parsons 1951 The Social System Taylor & Francis Abingdon
Pismenny, Arina, Arina Pismenny, Gen Eickers, and Jesse Prinz. 2024. "Emotional Injustice." *Ergo an Open Access Journal of Philosophy* 11 (0). https://doi.org/10.3998/ergo.5711.
Plunkett, David. 2021. "Debate: Anger, Fitting Attitudes, and Srinivasan's Category of 'Affective Injustice'*." *Journal of Political Philosophy* 29 (1): 117–31. https://doi.org/10.1111/jopp.12241.
Rawls, John. 1997. *A Theory of Justice*. 22. print. Cambridge, Mass: Harvard Univ. Press.
Reed, Pamela G., and Cindy J. Rishel. 2015. "Epistemic Injustice and Nurse Moral Distress: Perspective for Policy Development". *Nursing Science Quarterly* 28: 241–244. https://doi.org/10.1177/0894318415585634.
Ryan, Emer, Kevin Hore, Jessica Power, and Tracy Jackson. 2023. "The Relationship between Physician Burnout and Depression, Anxiety, Suicidality and Substance Abuse: A Mixed Methods Systematic Review." *Frontiers in Public Health* 11 (March): 1133484. https://doi.org/10.3389/fpubh.2023.1133484.
Shah, Megha K., Nikhila Gandrakota, Jeannie P. Cimiotti, Neena Ghose, Miranda Moore, and Mohammed K. Ali. 2021. "Prevalence of and Factors Associated With Nurse Burnout in the US." *JAMA Network Open* 4 (2): e2036469. https://doi.org/10.1001/jamanetworkopen.2020.36469.
Spranzi, Marta. 2018. *Le Travail de l'éthique: Décision Clinique et Intuitions Morales*. Psy—Théories, Débats, Synthèse. Sprimont: Pierre Mardaga Editeur.
Srinivasan, Amia. 2018. "The Aptness of Anger." *Journal of Political Philosophy* 26 (2): 123–44. https://doi.org/10.1111/jopp.12130.
Stockdale, Katie. 2024. "(Why) Do We Need a Theory of Affective Injustice?" *Philosophical Topics*.
Wardrope, Alistair. 2015. "Medicalization and Epistemic Injustice." *Medicine, Health Care and Philosophy* 18 (3): 341–52. https://doi.org/10.1007/s11019-014-9608-3.
Weilenmann, Sonja, Ulrich Schnyder, Brian Parkinson, Claudio Corda, Roland von Känel, and Monique C. Pfaltz. 2018. "Emotion Transfer, Emotion Regulation, and Empathy-Related Processes in Physician-Patient Interactions and Their Association With Physician Well-Being: A Theoretical Model." *Frontiers in Psychiatry* 9 (August): 389. https://doi.org/10.3389/fpsyt.2018.00389.

Whitney, Shiloh. 2018. "Affective Intentionality and Affective Injustice: Merleau-Ponty and Fanon on the Body Schema as a Theory of Affect." *The Southern Journal of Philosophy* 56 (4): 488–515. https://doi.org/10.1111/sjp.12307.

Chapter 2
Healthcare Providers and Affective Injustice

Abstract In this chapter, I will investigate the reasons why healthcare providers' affects have been downgraded in medical training and practice and the problems this causes. While most of the discussion in this chapter will focus on doctors, who face significant pressure to silence and suppress their affects, I will also consider if nurses are a special case and what specific affective harms they may face in their professions. I will first discuss the reasons that affects have been disregarded in medical training and practice. Then I will show how affects, contrary to popular belief, may actually be powerful motivating factors that may help healthcare providers to excel in their professions. I will also discuss how emotional suppression strategies are rarely successful and how they may actually lead to harm for professionals. Finally, moving from the individual to the collective level, I will discuss the ways in which affective injustice may lead to poorer quality care, including how it may lead to medical errors and compromise patient safety. This chapter is therefore focused on both the affective harms faced by individuals in the practice of their professions, but also the harms from the public health perspective from suppressing affects in healthcare service and delivery.

Keywords Healthcare providers · Doctors · Emotional regulation · Affective harms · Silencing

2.1 Introduction

Affects are shaped by our social worlds. The means by which we interpret, make sense of, and react to experiences depends both on how we develop and acquire affects but also how others receive and evaluate them. However, in some situations, they are not well received and are considered a barrier to effective communication. This occurs frequently in the workplace, where professionals are largely expected to communicate their expertise in a non-personal way, which often involves suppressing them in order to appear professional or objective.

B. Bogaert, *Affective Injustice in Healthcare*, The International Library of Bioethics 112, https://doi.org/10.1007/978-3-031-94378-2_2

In healthcare, affects have long been devalued in the push toward scientific objectivity and evidence-based practices. Healthcare professionals however do not do "just any job": they deal with life and death situations that have dramatic effects for persons under their care. These experiences and how they feel about them cannot be easily separated. They have to deal with many feelings as they navigate relationships with patients and their families, to find ways to handle uncertainties of care that can have dramatic effects on the life—and death—of their patients. Healthcare is truly an emotional experience, one in which professionals have to learn to handle their patients' affects but also to navigate their own.

Despite these realities, healthcare professionals are trained to distrust and discourage their own affects and those of their colleagues. This distrust is based on several ingrained ideas that are learned throughout training and honed in daily practice. Stockdale (2024) calls these *affective norms*, or socially accepted standards for emotional experience and expression. I will briefly describe the affective norms and ideas that have led to their mistrust in healthcare training and practice, before turning to its effects on healthcare providers and care delivery. Table 2.1 summarizes these arguments and their problems.

The first argument is that affects do not help with clinical judgement as they impede rational decision making. To "focus on the science" by remaining affectively neutral is supposed to help them to avoid being muddled in their judgement and enquiry. Doctors in particular are socialized in training to believe that affects may interfere with their scientific objectivity. Students therefore learn that a certain amount of *emotional distancing* will be necessary to be good scientists and clinicians. They

Table 2.1 Affective ideas and norms in healthcare training and practice

Arguments against affective expression	Problems
1. Affects negatively influence *rational decision-making*	Humans cannot make good decisions without affects; also ignores the healthcare provider's affective needs
2. *Ethical perspective* as we naturally feel positive/negative feelings toward certain patients, suppressing affects is a way to treat every patient the same (equity and non-discrimination)	Not paying attention to affects prevents being alerted to unconscious bias and other factors that may help work toward greater equity and quality care
3. Suppressing affects is a way to ensure the *epistemic authority* of the doctor	Suppression may negatively influence effective patient-doctor communication, shared decision-making, or quality of care
4. Affects impede *interprofessional coordination* and lead away from patient-centered perspectives	Affects may alert to important team disfunctions; they may also show something important about care and relationships with patients
5. Affects can cause *harm* to healthcare providers and must be avoided for their own protection (for instance to avoid having their work erode on their personal lives)	Suppression could mean that healthcare providers have no resources to deal with these experiences, thus doing them more harm in the long-term

are encouraged to find ways to manage—to keep at a distance—their affects. This happens at first in interactions between students and professors, where they are discouraged from showing their emotions, and then branches out in their engagement with patients.

From an ethical perspective, pushing affects aside is also seen as a way to ensure that the healthcare provider treats each patient the same. As there is a natural tendency toward favoring those that resemble us, whether in age or personality, or because they remind us of ourselves or of someone to whom we are fond (such as our grandmothers), healthcare providers inevitably feel more toward certain patients and act in consequence. In some circumstances, this may lead toward favorable treatment that may benefit that particular patient (such as when the doctor is engaged in doing their upmost to help them), but it will be unfair and unequitable to others, as it means that some will benefit from more resources and expertise. However, while this kind of special attention may be favorable to an individual patient, it may also disserve them in certain situations. Consider doctor Danielle Ofri's feelings about her patient Julia, an undocumented immigrant, who was admitted to her hospital due to congestive heart failure (2014, p. 26). She says,

> It was a little bit like looking in a mirror. We were the same height and build, the same age – mid-30s – and both of us had young children at home. In another world, had we been friends, we could have easily shared clothing. But today it was me with the white coat and her with the death sentence.

Dr. Ofri's feelings about this patient helped her identify with her and to feel empathy. However, it also made it difficult to tell her patient the truth about her future. This was time sensitive for Julia as the inability to get a heart transplant due to her immigration status was a death sentence. Dr. Ofri's analysis of the situation led her to believe that she was too emotionally engaged to be fully transparent in her communication with her. While her book goes on to show the importance of what doctors feel in the practice of medicine, and how having feelings toward patients like Julia have made her a better doctor, this case is emblematic of the type of ethical quandary facing doctors. In the case of Ofri, due to differential treatment, this led her to avoid important discussions with her patient.

On the flip side, healthcare providers (consciously or unconsciously) often have negative feelings toward certain patients, which may lead to discriminatory practices. This problem is often described as patient profiling. By this term, I refer to cases in which the healthcare provider may unfairly judge another person based upon their race, age, gender, or appearance, resulting in differential treatment that can cause harm. Pamela Wible (2014), a family physician in Eugene, Oregon, launched a social media campaign in 2014 to gather patient stories of profiling, many of which can be understood as cases of epistemic injustice (Fricker 2009). Take for example the case of Lonnie Stone, who was profiled as a cocaine user and not believed or understood, leading him to receive suboptimal care. According to him,

> As a teen, I fractured my nose. Many sinus issues later, I consulted an ENT specialist. He insisted that I damaged my sinus passages by using cocaine. His assumptions caused me pain, humiliation, confusion, and anger. I repeatedly assured him that I wasn't a user. Two

> surgeries later, my septum was removed. Afterwards, he was so cruel as to continue his tirade about my cocaine use. As the gauze was being removed from my nose, I fainted. When I was roused, he insisted that I leave immediately showing no concern about whether I could even make it home safely. (Wible 2014)

The negative feelings of this physician toward his patient due to assumptions about his health behaviors led him to not believe him. Not only was this a case of testimonial injustice, but it also led to a conflictual relationship with him and sub-optimal care and communication. While many healthcare providers are not consciously aware of such prejudices, their feelings about certain patients are mistrusted for this reason. As we can see from these examples, from the perspective of equity and non-discrimination, affects may be problematic. The conscious attempt to put them aside may help ensure that there is no difference—and no discrimination—between patients due to the doctor's feelings about them.

The third reason that affects are devalued in healthcare practices is doctors believe they need to maintain a certain professional distance to establish and maintain their epistemic authority. By epistemic authority, I mean authority based upon knowledge. By virtue of their training and professional roles, healthcare providers (doctors in particular) are tasked with providing their expertise and are expected to recognized as such by patients, their colleagues, and management. It is believed that the way to demarcate this authority in relation to others—in particular the patient—is to stay professional/an expert by minimizing or suppressing their affects.

The fourth reason to distrust affects is for interprofessional coordination. While the push to discourage affects primarily concerns patients, it is also a process that takes place in interdisciplinary coordination. In these temporalities, professionals are encouraged not to show their affects to other professionals. This serves a two-fold purpose. First of all, it is the means by which they are recognized as professionals by others, thanks to (but also in spite of) hierarchies. In other words, the lack of affects is the way in which a professional may be recognized for their expertise. The second is in a perspective of care ethics, as it is believed regulating one's affects provides the means by which professionals may focus on the task at hand, namely their work toward their patient. Here again there is the presumption that affects will interfere in clinical judgement and decision-making, as it is believed they will necessarily bias the discussion or distance it from its purpose.

Finally, there is the argument that a necessary emotional distance from what happens on the job is needed. This stance is advocated "for their own protection." There is a belief that what happens in the workplace will erode on their personal lives and because of this, there is a need to regulate or even suppress their emotions, otherwise difficult experiences of care will come home. Because burn-out and compassion fatigue are important problems of the profession, allowing oneself to feel with patients and their families is believed to favor this erosion, in particular when healthcare providers cannot help but be affected by difficult experiences with patients and families and their own inabilities to help them.

As we can see, the fact that affects have been silenced and discouraged in healthcare practices and interprofessional coordination has a certain logic and is even turned toward an ethics of care. Because of it, healthcare professionals try to do their best to hide or suppress their affects in front of patients, families, and their colleagues. These emotional regulation strategies are understandable. As we have seen, having strong emotions toward a particular patient may indeed interfere in some cases with clinical judgement and patient communication. While it can motivate them to search for the best possible treatment toward a person with whom they empathize, it could also lead to less quality care for others, and in particular those for whom they feel fear, anger, or disgust. The attempt to hide or push aside their affects can therefore be seen an attempt to avoid mistreatment and epistemic injustice. In the same light, in the workplace, holding an interprofessional meeting in which anger toward the patient becomes the focus may help to relieve some tension, but it also blocks a necessary work toward finding the best solution for the patient if it overwhelms the discussion. We also cannot expect that professionals take on the emotional burden of each patient and we should question whether or not this is even helpful (or desired by) patients and families themselves. For all of these reasons, affects may indeed at times have deleterious effects on healthcare provision and coordination.

What I am interested in exploring in this chapter however are those situations in which downgrading the healthcare professionals' affects has either had a negative effect on them personally (as persons, as professionals) and those situations in which it leads to poorer quality care or less fruitful interprofessional communication. In this chapter, I will be particularly interested in two interlocking problems. The first is that healthcare professionals inevitably feel—and therefore we need to deal with these feelings. As I will show, contrary to popular belief, they can be powerful motivating factors to help healthcare providers excel in their professions. From a more negative perspective, their suppression may also cause a number of affective harms. As I will show, these strategies are rarely successful and have detrimental effects on healthcare provider well-being. In the second place, there is a need to understand the institutional perspective and investigate what happens when healthcare professionals suppress their emotions and what effect this may have on healthcare quality. As I will show, the emotional suppression strategies advocated for with the intention of fostering better interprofessional or patient communication often have the opposite effect intended and may lead to poorer quality care.

While some scholars have argued that the emotional turn has come to healthcare training and practice thanks to movements such as person-centered care, as I will show in this chapter, distrust for affects is still standard practice and the reality of how medical care is being taught and delivered. My goal in this chapter is therefore to advocate that we need to take seriously situations in which downgrading or ignoring affects may be putting both patients and healthcare providers into harm's way.

2.2 The Affective Harms of Suppressing Emotions

Having clarified the reason affects have a difficult place, let us now try to better understand the place for healthcare provider's affects in medical training and practice. In the first place, I will attempt to turn the tide and show the positive side of showing and accepting affects, and notably how they can serve as powerful motivating factors for professionals when taken seriously. I will then turn to the affective harms that can occur for the person when affects are suppressed or controlled. In order to clarify this subject, I will show how emotional regulation and control happens in practice and the affective harms it causes.

2.2.1 *Affects as Powerful Motivating Factors*

In this section, I will argue that affects may serve as powerful motivating factors for professionals—when valued and accepted—and how may both help ensure quality care and provide strong personal motivation. I have chosen to discuss here some examples from early ethnographic studies in sociology. These works are still cited today in reference to the emotional socialization process which happens in medical training, and notably how students learn to separate their affects from their professions (see Chapter 8 for more discussion on this). However, less attention has been given to what these studies say about the powerful motivating force of affects for future doctors. This is surprising, as these studies are full of such examples. I will first of all give an example from Renée Fox's work before turning to Howard Becker's studies on student perspectives.

To start with, I will discuss Renée Fox seminal ethnographic work (1998), which consisted in following a group of patients and clinician scientists of the so-called "Metabolic Group." This research group hoped to learn more about the metabolism of the human body and the nature of various metabolic and endocrine diseases. They conducted early clinical studies with patients facing severe chronic diseases and often had to deal with the death of their patients—due to the severity of their diseases, but also because of their research interventions. Fox's study focused on how these clinician-scientists dealt with uncertainty in research and care practice and discussed the ethical dilemmas they faced in these early studies. Because of the long-term nature of their research, Fox documents how these clinician-scientists inevitably developed strong relationships with their patients. Patients often returned to the research center several times—sometimes even for several years—to participate in their research studies. In her description of the work of these scientists, what is striking is the lack of evidence of "detached concern."[1] In her study, we see an environment where affects toward their patients are valued. Because these clinician-scientists felt *with* and *about* their patients, they were also strongly motivated to work toward both scientific excellence and quality care. She describes the activities of these physician-scientists by saying,

[1] This term, developed by Fox, refers to the emotional socialization process in which future doctors learned to detach themselves from their affects in order to become professionals. See Chapter 8 for more discussion on this term.

> What emerges from such an observation is a picture of a group of medical scientists who actually thinking, feel, and act in ways which are considerably less 'cool, impartial, detached' and rational than would be the case if their attitudes, values, and behavior patterns conformed perfectly to some of the ideal norms of science....in their laboratory and conference room...they expressed a great deal of emotion about their shared difficulties...in these respects, the attitudes...of the research physicians of the Metabolic Group were not in keeping with the graven, aloof, logical-rational ideal image of scientists widely accepted as an accurate portrayal of how they actually feel and act (Fox 1998, p. 243).

While Fox describes the pull toward being "cool and impartial" clinical scientists, we also see that the clinician-scientists in her study—convinced by and developing evidence-based practices—were also unable to stay emotionally detached from their patients. She describes their relationships with their patients as practically "friends of the family," as they inevitably started to develop personal relationships with them, including ongoing mail correspondence beyond the research study about events in their personal lives.

While this kind of relationship between clinical scientists and their patients may be rarer today, it is also an example of how developing feelings—and accepting these feelings both in the team and toward their patients—was not an impediment to their being good professionals. These physician-scientists cared about their work. This was because they wanted to help these individual patients, but also because they were ambitious researchers who wanted to make a difference in their profession. These two ambitions were not mutually exclusive. It even made their research—which was full of uncertainty and risk—bearable. While Fox claimed that the actions and behaviors of the Metabolic Group were different than how doctors usually practiced medicine during this time, this example clearly shows how developing feelings toward their patients may help—and not necessarily impede—quality care for patients. In the case of the Metabolic Group, it even helped them to excel in their research work.

In a complementary way, Becker et al.'s study (2004) showed how students developed various perspectives to succeed in their training. They make the link in particular to their feelings and developing a sense of responsibility to ensure that they do their best for their patient. Indeed, their descriptions of the fear and anxiety medical students felt at the end of their training when realizing that they did not (yet) have enough knowledge was a strong motivating factor for them to invest in further training. The researchers show that nearing their internship and resident years, students were anxious about their lack of knowledge. They were fearful their ignorance would lead them to harm their patients (or at least provide suboptimal care). However, these affects did not stop them from acting. They were instead powerful motivating factors that led them to practical strategies to cope with their difficulties. This included choosing a certain type of hospital for their internship, such as choosing a general hospital to get experience of a diverse number of cases. It also led some to decide to work in a hospital group to be able to benefit from the experience of their peers. Others decided to choose a specialty to ensure they could develop sufficient knowledge in one area, as they believed that being a generalist was too much of a burden as they had to have extensive knowledge of all possible diseases. While these decisions were diverse, what we can see from their study is that these affects were not a barrier. They were instead part of a decision-making process turned toward the needs of their patients. Ignoring their fears and anxieties might have led to decisions they would later regret. Taking them seriously led to choices that were realistic and turned toward an ethics of care. They were also powerful motivating factors to stay humble and learn the most they could during their training.

These classic ethnographic studies give several useful examples of how affects may actually serve professionals to become better doctors and to work in an ethics of care. As we can see from these examples, feelings toward their patients helped them to make decisions that were in their best interests (further training in Becker's case, scientific excellence in Fox's case). They also helped them to recognize the limits of their knowledge and to seek out more training. Thirdly, they encouraged them to work toward scientific excellence. They were concerned for their patients and they felt about and toward them, and this helped them be motivated to act in the patient's interest and to excel in their professions.

When taken seriously as forms of knowledge, doctors were able to mobilize their affects—and what was behind them—to think about what was salient from their perspective. They understood them as related to how they see themselves as professionals and their ideals on providing good care. They are therefore not "just" an emotional reaction, but turned toward something that was important to them their professional roles. As we can see here, when affects are mobilized in this way, they are a facilitating factor rather than an impediment to providing quality care and improving decision-making. This shows the positive side of taking seriously affects. Let us now explore how ignoring them may lead to specific harms.

2.2.2 *The Negative Effects of Emotional Regulation and Control*

In this section, I will clarify the emotional regulation strategies used by healthcare providers to downgrade, defer attention from, or suppress the affects they experience. I will rely here on the concept of emotional regulation—developed in psychology but also in the emerging affective injustice literature—to better understand how healthcare providers have tried to "regulate" or "control" their affects and the harms this may cause. By emotional regulation, I refer to, "the processes by which individuals influence which emotions they have, when they have them, and how they experience and express them (Archer and Mills 2019, p. 79 quoting James Gross 1998, p. 275). Emotional regulation is not an innate capability; it is a learned and dynamic process that enables us to live with others and adapt to them. It is a social skill which enables us to participate in the social world, but it can also go awry, as we will see.

It must be stated before continuing that I am not suggesting that emotional regulation by healthcare providers is necessarily negative. There are several reasons to regulate one's affects in the practice of care. Studies on emotional regulation have shown that generally, it enables a person to adjust the expression of their emotions to the demands of the environment and to avoid one's negative emotions from interfering with their functioning (Kadović et al. 2022). It can also help persons to respond to stressful demands and emotional experiences in an adaptative and flexible way. Emotional regulation is also a cognitive process: it can help the person to analyze the reasons causing the emotion, the choice of a reaction, and if necessary engage a conscious effort to postpone or suppress it. From this standpoint, emotional regulation strategies can help the person feel a sense of control over their emotions and to cultivate the good emotion at the good time. In the context of healthcare, successful emotional regulation has been shown crucial to enable well-being and prevent burnout and compassion fatigue in healthcare professionals (Weilenmann et al. 2018). Due to stressors such as breaking bad news or dealing with aggressive patients, regulating emotions toward the patient or family (such as suppressing one's anger in order not to aggravate the person's aggressivity) may allow healthcare providers to provide quality patient care, in particular in difficult and stressful situations. Furthermore, too much emotional investment can lead to distress, compassion fatigue, and burnout, such as when healthcare professionals are overburdened by the strains and experiences of their professions.

However, despite the advantages of emotional regulation, it can also cause harm. In this section, I will focus on the harms they may cause to healthcare professionals. These harms have (already) been conceptualized in the affective injustice literature by Archer and Mills (2019), following their extensive development in psychology. They include:

(1) Situational strategies (managing the situations one finds oneself in)
(2) Attentional strategies (managing what one attends to)
(3) Cognitive reappraisal strategies (managing the appraisal of one's situation)
(4) Response modulation strategies (managing one's behavior, including emotional suppression strategies)

In order to better understand these emotional regulation strategies in the context of healthcare and the harms it may cause, I will elaborate these four strategies via a common situation experienced in interprofessional work, namely that of a professional yelled at by another professional because they are stressed. This often occurs in situations of hierarchy, such as when a nurse is yelled at by doctor, be it because of an error, because they are not working fast enough, because they disagree about an aspect of patient care, or simply because they are having a bad day. In this situation, I will assume that the anger by the doctor is inappropriate and not based upon any fault on the nurse's side. This will help focus on situations in which the nurse may be justifiably wronged but has to choose how to react to the situation.

In terms of the nurse's perspective, they are first of all likely to be justifiably angry. Furthermore, we can sympathize with the nurse and understand that they may wish to show their anger in response to this aggression from the doctor. However, showing this emotion is a risk for the nurse, as it may further amplify the conflictual situation. Indeed, based upon our earlier discussions, we know that showing emotions may be seen as unprofessional, even if the doctor has shown their anger first. Showing anger is therefore unlikely to help the nurse be valued and for the wrong to be recognized and redressed. In this dilemma, what should the nurse do?

A first technique that may be employed by the nurse is a situational strategy. This involves avoiding places and situations that make us feel a certain way. For instance, if we feel oppressed by certain persons, we may seek to avoid these persons in order to avoid being negatively influenced by them. Situational strategies may also involve attempts to deliberately steer a conversation away from a certain topic when we cannot avoid the person. Therefore, the nurse who is yelled at by the doctor has several options. The first is to avoid the doctor or other persons who are aware of the situation. The second strategy may be to try and diffuse the situation by steering the conversation to another topic (with an uncertain result, as the doctor is "riled up").

The advantage of these strategies from the nurses' perspective is it may help the nurse to lessen their discomfort by taking them away from the persons oppressing them or even diffusing the situation by changing the topic. In this way, they will avoid being negatively influenced by them. The problem with this strategy however is that it may lead to avoidance of the problem. It also does little to change it. Indeed, these kinds of strategies may lead to poor communication among professionals in future situations. Such strategies can therefore be harmful both to the professional—who did not find an adequate way to deal with their justified emotions—but also to the patient who does not benefit from interdisciplinary coordination due to this communication breakdown. In this situation, the nurse has been harmed as an affective being (their emotional needs were dismissed) and the conflictual situation has not been resolved. Patients have also been potentially harmed from the communication breakdown in the team.

A second emotional regulation strategy the nurse may wish to engage in is attentional. It involves shifting one's attention from the object causing an undesirable emotion and/or directing one's focus elsewhere. By using these tactics, the person may be able to control their affects and the ways they react to it, thus helping to lessen negative and painful emotional responses. Archer and Mills (2019) give the example of women suffering sexist remarks in the workplace. By thinking about something else while they have to listen to sexist conversations, they find a way to lessen or prevent their anger. However, like situational management strategies, the problem with this tactic is that it makes it hard to challenge problematic behaviors and may harm the person's well-being. It is also a missed opportunity for them to consider in what ways and means the oppression operates and to evaluate the extent of the problem. As Archer and Mills state (2019, p. 83), "by choosing not to pay attention, one is depriving oneself of vital epistemic resources to understand, challenge, and survive." In my example of the doctor who yelled at the nurse, rather than showing their anger (which the nurse believes may cause problems for future relations), the nurse may instead try to focus on their professional tasks with other patients or what they will do after work. In the short run, while this may help to (immediately) defuse the situation, in the long run it does nothing to change the situation and will likely further impoverish the relationship between the two. The injustice of the situation is also that it is not acknowledged, which causes harm to the nurse by not giving them an outlet to express justified anger. It may also cause the oppressor (in this situation, the doctor) to think it is acceptable to act in such a way with other colleagues.

As we can see, both of these emotional regulation strategies may interfere not only with interpersonal communication but also have a negative effect on quality care. The third strategy of emotional regulation and control is just as problematic, if not worse. Cognitive reappraisal strategies involve the ways in which we seek to change our emotions, notably by changing how a person evaluates the situation provoking the emotion. This may involve creating distance from the situation in one's mind or deciding to evaluate it in another way. In psychology, cognitive reappraisal has been found effective to control the experience and expression of negative emotions. It has direct physical benefits, such as lowering blood pressure in response to provocation (Memedovic et al. 2010). However, the obvious problem with this strategy is that it leads to a new evaluation of a situation that may not be a wrong in the first place. For instance, the nurse who was yelled at by the doctor in the previous example has been oppressed by them. A way the nurse may choose to deal with it is by reevaluating and deciding somehow that the situation is their fault. However, once again the nurse has little to change the actual problem, causing them affective harm, in particular as it has not been adequately acknowledged either by the oppressor or the oppressed. It can also cause epistemic harms, as the person may lose confidence in what they believe or know.

The final strategy discussed by Archer and Mills is response modulation, which looks at how one responds to the emotion experienced by seeking to directly influence how we react. Although this may lead to some dangerous behaviors (such as taking drugs or using alcohol in order to change how we feel), it may also lead to behaviors that can be positive to one's health (such as doing sports in order to feel better). In the context of healthcare, this strategy can be particularly useful to avoid having a situation take over one's personal life. If we return again to our example of the nurse, in order to avoid having this situation influence their personal life, they may decide to deal with this oppression by engaging in more positive experiences, such having lunch with their friend or exercising during lunchtime. This may help them to modulate their response to the situation and prevent the oppressive situation from taking over.

However, when the response modulation takes the form of *emotional suppression*, it can become particularly problematic for the person experiencing it. This concept refers to situations in which we prevent ourselves from expressing an emotion, such as trying to control facial muscles or one's voice to avoid having the other person witness our emotional response. If we come once again to the example of the nurse, if they attempt to deal with the situation by controlling their facial muscles (therefore actively seeking not to show anger at being treated unjustly), this can cause them important affective harms. Studies have shown that emotional suppression leads to a decrease in the positive emotions one experiences, but does not necessarily help to change our experiences of negative emotions. It has even been found to lead to impaired cognitive functioning and increased risk of cardiovascular diseases (John and Gross 2004). In addition, suppressing one's emotions does not mean that they necessarily die out: as we know from persons close to us who may "lash" out when they experience problems in the workplace, it may resurface in inappropriate situations and be redirected toward persons whom we had no intention of hurting. Thus, while this kind of emotional regulation strategies may help in the immediate to diffuse the situation, it does little to avoid the affective harms experienced by the person in their overall lives. What is even worse is that the harm may have expanded and transform into lashing out on their family members, making the problem even larger and having workplace experiences intrude on their personal lives.

To conclude, these four forms of emotional regulation strategies may help in the short term, but in the long-term cause harm to the person. It also does little to change the conflictual problem in the first place. In this light, such strategies have the potential to negatively affect healthcare providers' well-being but also to further impoverish relationships with patients and co-workers. As we can see, these forms of emotional regulation strategies largely involve intrinsic emotional regulation strategies, or in other words ways in which persons actively seek to change how they react to, evaluate, or control their emotions. The intrinsic emotional control/regulation strategies that healthcare providers use plays an important part in how affective injustice happens in practice. I call this practice *affective self-silencing*, as it involves in most cases silencing the emotion in both the person experiencing it and to outsiders (including the oppressor). As we have seen, healthcare professionals attempt to silence their emotions in a logic that they need to do so in order to remain professional and/or to diffuse conflict. By not valuing their own affects, however, they are not paying attention to their emotional needs, a problem which can have spill-over effects not only on patient care but also in their personal lives.

While many forms of emotional regulation used by healthcare providers appear to be intrinsic, the extrinsic nature of emotional regulation in healthcare is not to be forgotten. These extrinsic control strategies are not necessarily explicit, but find a way to be communicated through actions and behaviors. Going back to the example of the nurse and the doctor, this professional was not told by their co-workers that they did not have a right to express their justified anger; however, hospital hierarchies and their past experiences made it obvious that it would not be appropriate to show their emotion. Furthermore, because they are based on ideals which many healthcare professionals share—that emotions are unprofessional—there is little real attempt to change the situation and encourage their expression. Yet as we have seen from the example of the doctor and the nurse, emotional regulation strategies often have the opposite effect of what they have intended: they spill out into the professional's personal lives; they lead to less quality care; they encourage communication breakdowns. Having explored the dangers of emotional regulation strategies, it will now be necessary to explore in more depth the affective harms healthcare professionals experience.

2.2.3 *Affective Harms*

This section will look specifically at the affective harms that may come out of strategies of emotional regulation. As we have seen, healthcare providers work in an environment full of affects, and yet they are expected either not to have or at least not to show these emotions to others, notably because it is believed that it would interfere with quality patient care and interdisciplinary collaboration. The recurrent idea that "emotions make one unprofessional" or that "we need to get beyond emotions" to work together in teams reinforces a social expectation that healthcare providers be strong, resilient "machines" or even "soldiers" that can take any difficulty with stride. This means that healthcare providers who do not live up to these ideals may feel inadequate. This may exacerbate distressing emotions, including feeling shame for feeling an emotion in the first place. This is the first problem and it is a big one.

The second problem is emotional disorientation. When we discourage affective expression in healthcare professionals, when we devalue them and question their worth, healthcare professionals may develop doubt about what they value and experience. This is because downgrading the value of emotions means that they are disorientated (distanced or turned away) from their object, an object which may be of considerable importance to the person in their professional (and personal) lives. For instance, in the example of the nurse discussed above, when this person does not give sufficient attention to why they are feeling this way, they may also be discounting what they value and their place in the team. The ambiguous place that emotions currently have in medical training and in healthcare institutions have led to what Gallegos (2021, p. 10) has referred to as *affect-related hermeneutical injustice*. This type of affective injustice includes situations that strongly encourage self-silencing of affects. Other than causing individual harm, this also leads to a conceptual lacuna in the collective imaginary: it means that we do not pay sufficient attention to healthcare providers' affective experiences and what this may say about what they value in their professions, leading to affect-related hermeneutical injustice. In other words, it creates an environment in which the society's ability to interpret their own and others' affective experiences is unfairly constrained or undermined (Gallegos 2021).

In these situations of affective injustice, there are both affective and epistemic harms that may occur. For the individual person, affective self-silencing means that they are missing out on understandings of themselves as professionals (but also more widely as persons). As seen in Fox's example of the Metabolic Group, the affects the clinician-researchers felt toward the patient was a source of strain—but also helped them to focus their attention and energy on why they did what they did. It motivated them in their professions and helped them to question the limits of their actions and the effects of it on their patients. It also helped them to consolidate as a research group and find ways to deal with these affects collectively, rather than ignoring them. What they felt about themselves and their patients was a source of knowledge and helped them to recognize themselves in the wider scheme of their lives and professions. With the Metabolic Group, we are far from the ideal of healthcare professionals as impersonal, unaffected soldiers—and closer toward finally recognizing them as human beings, in their capabilities and vulnerabilities.

Furthermore, while the discussion thus far has focused on the individual epistemic and affective harms that the person as an individual may experience, affective self-silencing of the kind described here also represents an epistemic harm for the institution. When institutions ignore the affects of their healthcare workers, they are also missing out on valuable information that may help them to better respond and provide a caring environment for healthcare workers—and in return for their patients. The harm of not taking seriously a healthcare provider's affects—or actively seeking to silence them—is an epistemic injustice. It involves wronging the healthcare provider in their capacity as a knower, as the knowledge that is made available and visible through emotional expression is not taken seriously or understood. However, before continuing our discussion, let us take a moment to understand if nurses and other paramedical professionals face specific affective harms in their roles as nurses.

2.2.4 Nurses: A Special Case?

Before turning to the case studies, which will deal with how emotional regulation affects quality of care, I would like to take a moment to discuss the specificity of the nursing profession in terms of affective injustice. I have purposely used the term "healthcare providers" in this book rather than "doctors" or "clinicians" in order to include the needs of nurses and other paramedical professionals in my analysis. The discussion thus far has already exposed how nurses may be subject to affective injustice in function of hospital hierarchies. However, is there any particular thing to say about how nurses self-silence and the reasons for this?

First of all, it should be recognized that nurses have very different relationships to their patients than clinicians. What is valued in these relationships, and how and in what ways affects may be expressed, changes how affects are viewed, recognized, or understood by these professionals. Nurses have more direct and prolonged contact with patients than doctors (at least in the hospital context), which means that they are naturally closer to the patient. In addition, the nursing profession in the past has specifically valued affects due to the history of the nursing profession and in particular a religious past that emphasized love toward their patients. While most nursing professionals would not necessarily use this word to discuss their relationships with patients today, what is clear is that suppressing their affects is not necessarily a nursing ideal. Many nurses today value an emotional commitment to their patients as part of their profession and as essential to care quality (Font-Jimenez et al. 2020). Yet because of this, they are also more exposed to emotional demands and the risk of burn-out and compassion fatigue. Like doctors, nurses must (also) find a balance between emotional expression (including but not exclusive to empathy with and toward their patient) and emotional regulation strategies that enable them to protect themselves from burn-out and difficult care situations. They are also in need of safe spaces to deal with and discuss their emotions with other nurses and healthcare professionals.

Research has shown that emotion regulation strategies employed by nurses include trying to be positive, disconnecting, taking a protective shield, and taking distance. They also involve emotional suppression strategies, such as when nurses are expected to display their "professional face" while at work. This "display rule" means that they are expected to mask or control their negative emotions. It also involves an expectation that they display positive emotions that may not always been genuine (Fasbinder et al. 2020). These kind of emotional suppression strategies may be particularly harmful to nurses, as it is not just a question of trying to suppress one's emotions, but of physically changing the emotional expression itself. As Font-Jimenez et al. (2020) note, the risk of such strategies are that it may lead to a distant attitude with the patient, higher rates of stress and burnout, and feelings of being overwhelmed, as it remains difficult to conform to these ideals in practice.

In addition, I stated at the beginning of this section that nursing ideals and their perspectives on affects are changing. This is part of a wider movement across the healthcare profession that nurses have not escaped: the expectation that they put aside their emotions to be "professional." In movements that have sought to better value the profession of nursing, in particular in the context of interdisciplinary discussions, this push toward affective neutrality (Parsons 1951) has also played a role in silencing the affects of nurses to help guide healthcare practice. Furthermore, even if we accept that the nursing profession values affects, we also have to take into account the reality of a profession which is increasingly technical (Jiménez-Herrera et al. 2020) and subject to increased stress and less direct and prolonged contact with patients due to budget cuts. These factors are affecting nurses' capabilities to build relationships with their patients and in particular to "feel with them" in their experiences of care.

Within increasingly degrading working conditions, nurses are also not necessarily the persons making care decisions. They are often seen as "implementers" rather than decision-makers. Despite attempts toward greater interprofessional deliberation, hierarchies of knowledge and authority continue to prevent interprofessional decision-making (Price et al. 2014). Nurses enact decisions made by others, whilst not always being part of the decision-making process. While this may relieve some negative feelings, in difficult or conflictual situations in which they did not necessarily agree with the decision taken, it may lead to situations of epistemic uncertainty and feelings of powerlessness, all of which have been associated with moral distress (Reed and Rishel 2015). Moral distress has been defined as the psychological distress of being in a situation in which one is constrained from acting on what one thinks to be right (Jameton 1984). Nurses are in particular affected by moral distress, notably because they have more direct contact with their patients but also have less chance to act on what they think is right.

Moral distress is important to the affective injustice discussion in healthcare, as it means that nurses may feel certain distressing affects (frustration, anger, sadness) about decisions made and not be able to exercise or make visible these emotions, which further increases the affective harms they may experience. These situations take place in the context of patient care and in interprofessional coordination, where the nurse is called upon for their testimony and expertise but in which they are not necessarily part of the final decision-making process. As these decisions ultimately affect their patients, and as these professionals value their personal interactions and relationships in particular, decisions about which they may disagree—but in which they cannot express their affects—may cause them even greater affective harm.

2.3 Affective Injustice and Poor Quality Care

It is clear from our discussion that affective injustice is a recurrent problem in healthcare and it may cause harms to the individual and to the institution. As we have seen, not addressing affects and their impact on care leads to affective and epistemic harms for the individual and the institution. It means that healthcare institutions are missing out on valuable information they need to take care of their workers but also not addressing other important factors in care. In this final section, I will focus on situations in which affective justice not only harms the individual, but may also lead to poor quality care and patient dissatisfaction. I have chosen two common experiences of healthcare to illustrate this point. The first will focus on patient communication and show how emotional regulation strategies may interfere with shared decision-making. The second is medical errors, and in this discussion, I will show the negative effects of affective self-silencing to care quality and institutional learning. Table 2.2 summarizes the discussion to follow.

Table 2.2 Emotional regulation strategies and care quality

Reason	Strategy	Effects on care quality
Uncertainty; treatment failure	Attentional strategies	Conflictual doctor-patient relationships; strategies of avoidance
	Response modulation strategies (esp. suppression)	Lack of transparency toward the patient; insufficient shared decision-making
Medical errors	Cognitive strategies; attentional strategies	Conflictual doctor-patient relationships; deterioration of the team; emotional distress for healthcare providers

2.3.1 *Affective Suppression in the Face of Uncertainty: The Patient Knows*

This case study[2] will look at how doctors regulate their emotions in the face of uncertainty and how patients evaluate these strategies. I will discuss in particular the emotional regulation strategies (attentional and response modulation strategies) used by doctors in cases of uncertainty and the fear they experience (and show) when faced with difficult situations in front of their patients. I will notably discuss how emotional regulation strategies may lead to poor relationships and poorer quality care from the patient perspective.

This case study comes from qualitative research (ethnography of consultations and interviews with healthcare providers and patients) conducted in a socio-medical center in France with persons with severe refractory epilepsy. Severe refractory epilepsy can be understood as recurring seizures that cannot be fully controlled by anti-seizure medication (ASMs). Not being able to control seizures is harmful to persons with epilepsy and their families and their well-being, as it can lead to cognitive limitations, behavioral problems, and stigma. From a medical (but also social perspective), it is therefore a priority to find solutions to reduce—or if possible—to completely stop seizures all together. Despite the fact that epilepsy is one of the oldest known diseases, it is also one in which there are many unknowns. A treatment that works on one patient may not work on another; a treatment that may work for some time may also stop working for some unknown reason. In the case of severe refractory epilepsy, in which patients both have severe seizures but are unable to find an appropriate treatment, patients often say that they are a "mystery" or "fascinating" as doctors do not know what to do. The only clinical solution at present is to try out different combination of medications and to see their effects on that individual patient, to find out what works best for them to reduce their seizures—or even to "find the lucky ticket," that combination of medications that may even stop seizures altogether. For both the patient and the neurologist, there is therefore considerable uncertainty (but also hope) in the shared decision-making process.

While both the doctor and the patient experience uncertainty in these situations, how the doctor decides to communicate it is particularly challenging for them. First of all, it puts into difficulty their epistemic authority, as they are not sure of what to do. It may cause them doubt about themselves and their professional abilities. Because of this, at times, in the face of uncertainty, doctors may feel fearful. They feel fear of appearing incompetent in front of their patients, but also fear of choosing a treatment plan that causes harm. Fear is the most frequent emotion experienced by healthcare professionals, along with sadness, anger and frustration (Silva and Carvalho 2016) and often stems from situations of uncertainty. The problem I will describe here is what happens when the doctor tries to hide their fear from the patient. I will argue that

[2] This case study was originally described in a paper which looked at the fear that patients and healthcare providers experience, using Martha Nussbaum's (2003) concept of emotions. I will reuse this case study here to explore how it may also be a case for affective injustice. For the original article see Bogaert (2020).

it leads to practices of emotional regulation largely consisting of avoidance strategies which are highly problematic from the patient's perspective.

However, a side note before continuing. In relation to the discussion of the difficulties for healthcare providers to acknowledge their own emotions, it is particularly telling that neurologists did openly talk about the word "fear" or "apprehension" when describing the difficulties of communicating uncertainty on the treatment plan to their patients, but they often acknowledged its expression in their colleagues ("not me, but others may feel fear"). In addition, in interviews, neurologists also talked about their colleagues' affects as negative, a factor which prohibited them from interdisciplinary team work and from achieving a high level of competence in their professions. Clearly, in this highly technical profession, emotions are rarely given a safe place for expression. Is it surprising that they sought to hide them from their patients?

What is interesting about this case study is that: (1) doctors have a hard time admitting to their own affects and seek to actively hide or suppress them; (2) they are unable to hide them from their patients. Indeed, patients specifically named the word "fear" to describe certain encounters with healthcare providers. One patient expressed it as the following, "it's the capacity to put oneself under scrutiny, that it not easy…it's about ego…but it's also fear to put oneself under scrutiny. Fear." From the patient perspective, the healthcare provider's fear was not the problem. It even made them feel closer to their doctor, as they also felt fear due to their unpredictable condition and the uncertainty of the treatment plan. The problem from their perspective was the emotional regulation strategies engaged in by their healthcare provider to hide their emotions, in particular strategies of avoidance.

Patients described these emotional regulation strategies from a visual perspective. For instance, when asked how they knew that their neurologists were scared, they often described non-verbal behaviors of avoidance. These behaviors included looking at the computer screen instead of them, not engaging in discussions about treatment options, not proposing a new treatment option, etc. Concretely for patient, this meant that shared decision making did not take place and it resulted in a certain mistrust of their care providers, as they seemed detached or disengaged from their patient. They also left feeling their doctor did not listen to them.

Although this evaluation of healthcare providers' affects by patients obviously has some limitations (we could ask for example if this is really fear or whether is this may be a misevaluation by the patient), let us take this evaluation seriously and understand its effects on patient care. First of all, from the doctor's side, there are attentional strategies designed to avoid the conflictual or distressing situation. When the neurologist chooses to look at the computer instead of the patient, they are shifting attention away from the problematic situation (the patient) to a more neutral object. If they were conscious of this, it might have been a deliberate strategy by them to focus their attention elsewhere in order to avoid showing their fear to the patient. While this is an understandable reaction in the face of uncertainty, the evaluations by patients also show that doctors cannot hide this fear from their patient: their non-verbal avoidance strategies let the patient know that they are scared—not necessarily of their patient, but of the problematic situation in which they cannot provide assistance. The problem with this strategy from the patient's perspective is that it leads to inadequate communication about the treatment plan—and does little to address how they themselves feel in these stressful situations. We can also see here that the doctors were engaging in emotional suppression strategies, such as putting on a "brave face" to hide their fear. However, by not admitting—or trying one's best to hide—this uncertainty, little discussion with the patient took place, including a discussion with the patient about their own fears.

This case study showed that such types of emotional regulation strategies did little for the patient and harmed patient-doctor communication and transparent discussions with their patient. This led patients to feel less comfortable discussing with their doctors and feelings of frustration about the lack of communication and discussion. We can see here how this may negatively influence care quality, as shared decision-making or honest discussion on remaining treatment options—as well as a discussion of what measures were acceptable to the patient—did not take place. Patients were left feeling less informed and less implicated in their care. This also did not help to increase their trust in their doctor.

2.3.2 *Medical Errors and Patient Safety*

Having understood how emotional regulation strategies can poorly influence patient communication, I will now investigate the effects of emotional suppression on quality care. In this case study, I will look at how medical errors induce negative feelings and how not taking them seriously can negatively influence patient communication and safety. I propose here that not dealing with medical errors involves both *cognitive reevaluation strategies* (managing how one appraises the situation, such as avoiding blame), and *attentional strategies* (such as avoiding talking about the situation). As we will see, without institutional support, healthcare providers will be encouraged to deal with such situations with emotional management strategies that will likely have a negative influence on care quality and may lead to further errors.

To begin the discussion, medical errors both involve those "near-misses" (which involve adverse consequences that could have happened but did not) and an adverse event (in which an error results in patient harm). An example of the first is the wrong drug being prescribed but not administered; whereas adverse events happen when the wrong drug was prescribed, administered, and resulted in an unfavorable outcome for the patient.

Both of these types of errors cause emotional distress for the healthcare provider. It is however likely that the greater the patient harm, the more distressing it will be for them (Harrison et al. 2015). However, in the case of near-misses, as the adverse event did not actually occur, it is unlikely to have been dealt with by the healthcare team or the healthcare institution, which may increase the healthcare provider's distress about it. In this section, I will argue that not addressing the affects lived by healthcare providers due to medical errors has both a negative impact on patient safety, but also on their relationships with patients. Given the emotional distress experienced following such adverse advents, healthcare providers have even been called its second victims.

Recognition of the emotional impact of error on healthcare professionals is increasing. Physicians report significant distress in the face of a medical error, in particular anxiety about future errors, loss of confidence, and reduced job satisfaction (Waterman et al. 2007). Psychological repercussions include shame, guilt, fear, panic, shock and humiliation immediately following the error as well as broader psychological responses such as self-doubt, loss of confidence and altered perspectives of patient and colleague relationships (Sirriyeh et al. 2010). In turn, negative outcomes include loss of professional reputation, reduced goodwill toward patients, and detachment from patients (Sirriyeh et al. 2010; Heyhoe et al. 2016).

While patients and families report a desire to have errors disclosed to them, healthcare professionals are still reluctant to do this. From their perspective, it means admitting fault and may degrade the healthcare provider's epistemic authority. The response of patients and families is also unpredictable in the face of error announcements. Healthcare providers may face anger, resentment, or even violence from patients and their families. In addition, healthcare providers themselves experience many negative feelings about medical errors—whether or not they are directly responsible for them. When the patient expresses anger or resentment in such situations, they are often left feeling even more hopeless. Studies have shown that physicians are more likely to be emotionally distressed when dissatisfied with error disclosure to patients, meaning that strategies need to be developed in healthcare institutions to deal with medical error and its effects on professionals (Waterman et al. 2007).

Research suggests doctors and nurses respond differently to medical errors, although it is distressing to both for different reasons. In nursing, the reflective approach championed in this practice may be a source of additional anguish after an error, as nurses reflect on their responsibility for the adverse event. On the other hand, as doctors prioritize professional loyalty and patient retention, they sometimes ignore or deny an error, or even collude with colleagues to cover it up (Harrison et al. 2015). One of the reasons errors may be difficult for doctors to accept is that medical culture reinforces an ideology of internal infallibility which has been shown to create profound self-doubt and guilt in medical students when they do not live up to its ideals. This idea also further reinforces the idea of healthcare providers as "soldiers."

While much attention has been given to the negative emotions felt by healthcare professionals, less have been shown about what happens when they are recognized and openly discussed. When error disclosure is done in a safe environment, it can elicit positive emotions and behaviors important for an ethics of care, such as increased assertiveness, confidence, and improved relationships with coworkers (Sirriyeh et al. 2010). Research has shown that openly acknowledging the error helps the team and institution to work toward improvements in care practices, including changes to practices and corrective actions within the department and in the hospital. In reverse, such situations burden healthcare professionals when staff consider the institutional handing of the error to be poor (Sirriyeh et al. 2010).

These studies show it is worth supporting healthcare professionals both before, during, and after the error disclosure and that this means taking seriously and addressing the affects that arise, including when the error is discovered, when they make the error disclosure to families and other coworkers, and in the aftermath, when they have to live with the adverse impact of the error. By taking seriously the affects lived and experienced, it can also be a key learning moment for the team and the institution, as it will enhance their abilities to openly discuss the problems leading to the error and engage in realistic discussions of how to reduce their incidence in the future.

These two case studies are but two of a larger debate that needs to happen to better understand how suppressing, ignoring, or downgrading affects may influence care quality. They have also shown how emotional regulation strategies can be harmful to patient communication and quality of care. This is all the more ironic, as what motivates emotional regulation strategies in the first place is often a desire to provide the best quality of care. Clearly, healthcare providers' affects need to be taken more seriously and safe spaces found for them, and this includes accepting they experience them in the first place.

2.4 Conclusion and Next Steps

This chapter sought to provide a nuanced perspective of when emotional regulation strategies may help or hinder patient care and interprofessional coordination. It has shown the affective harms for healthcare workers of ignoring or suppressing emotions in some key situations, and the valuable information the healthcare institution may be losing out by ignoring them. If an emotional turn has arrived in the social and human sciences, it has yet to become a reality in healthcare practice. To foster greater attention and recognition of the affects of healthcare providers, there will be a need going forward to put into place affective resources, opportunities, freedoms, and recognition to take better care of healthcare providers. This is the purpose of Part II of this book. However, before that, it will be necessary to better understand when patients and families are subject to affective injustice and the harms this causes them.

References

Archer, Alfred, Georgina Mills, and University of Arkansas Press. 2019. "Anger, Affective Injustice, and Emotion Regulation." *Philosophical Topics* 47 (2): 75–94. https://doi.org/10.5840/philtopics201947216.

Becker, Howard S., Blanche Geer, Everett C. Hughes, and Anselm L. Strauss. 2004. *Boys in White: Student Culture in Medical School*. Fac-sim Ed. New Brunswick London: Transaction.

Bogaert, Brenda. 2020. "Untangling Fear and Eudaimonia in the Healthcare Provider-Patient Relationship." *Medicine, Health Care and Philosophy* 23 (3): 457–69. https://doi.org/10.1007/s11019-020-09956-1.

Fasbinder, Adrienne, Kimberly Shidler, and Meriam Caboral-Stevens. 2020. "A Concept Analysis: Emotional Regulation of Nurses." *Nursing Forum* 55 (2): 118–27. https://doi.org/10.1111/nuf.12405.

Font-Jimenez, Isabel, Laura Ortega-Sanz, Maria Sagrario Acebedo-Uridales, Maria Jesus Aguaron-Garcia, Inmaculada deMolina-Fernández, and Maria F. Jiménez-Herrera. 2020. "Nurses' Emotions on Care Relationship: A Qualitative Study." *Journal of Nursing Management* 28 (8): 2247–56. https://doi.org/10.1111/jonm.12934.

Fox, Renée C. 1998. *Experiment Perilous: Physicians and Patients Facing the Unknown.* New Brunswick, N.J., U.S.A: Transaction Publishers.

Fricker, Miranda. 2009. *Epistemic Injustice: Power and the Ethics of Knowing.* 1. publ. in paperback. Oxford: Oxford University Press.

Gallegos, Francisco. 2021. "Affective Injustice and Fundamental Affective Goods." *Journal of Social Philosophy* n/a (n/a). https://doi.org/10.1111/josp.12428.

Gross, James J. 1998. "The Emerging Field of Emotion Regulation: An Integrative Review." *Review of General Psychology* 2 (3): 271–99. https://doi.org/10.1037/1089-2680.2.3.271.

Harrison, Reema, Rebecca Lawton, Jessica Perlo, Peter Gardner, Gerry Armitage, and Jo Shapiro. 2015. "Emotion and Coping in the Aftermath of Medical Error: A Cross-Country Exploration." *Journal of Patient Safety* 11 (1): 28–35. https://doi.org/10.1097/PTS.0b013e3182979b6f.

Heyhoe, Jane, Yvonne Birks, Reema Harrison, Jane K O'Hara, Alison Cracknell, and Rebecca Lawton. 2016. "The Role of Emotion in Patient Safety: Are We Brave Enough to Scratch beneath the Surface?" *Journal of the Royal Society of Medicine* 109 (2): 52–58. https://doi.org/10.1177/0141076815620614.

Jameton, Andrew. 1984. "Prentice-Hall Series in the Philosophy of Medicine". In Nursing Practice: The Ethical Issues. Englewood Cliffs, N.J: Prentice-Hall.

Jiménez-Herrera, María F., Mireia Llauradó-Serra, Sagrario Acebedo-Urdiales, Leticia Bazo-Hernández, Isabel Font-Jiménez, and Christer Axelsson. 2020. "Emotions and Feelings in Critical and Emergency Caring Situations: A Qualitative Study." *BMC Nursing* 19 (1): 60. https://doi.org/10.1186/s12912-020-00438-6.

John, Oliver P., and James J. Gross. 2004. "Healthy and Unhealthy Emotion Regulation: Personality Processes, Individual Differences, and Life Span Development." *Journal of Personality* 72 (6): 1301–34. https://doi.org/10.1111/j.1467-6494.2004.00298.x.

Kadović, Marija, Štefica Mikšić, and Robert Lovrić. 2022. "Ability of Emotional Regulation and Control as a Stress Predictor in Healthcare Professionals." *International Journal of Environmental Research and Public Health* 20 (1): 541. https://doi.org/10.3390/ijerph20010541

Memedovic, S., Grisham, J. R., Denson, T. F., & Moulds, M. L. (2010). The effects of trait reappraisal and suppression on anger and blood pressure in response to provocation. *Journal of Research in Personality, 44*(4), 540–543. https://doi.org/10.1016/j.jrp.2010.05.002

Nussbaum, Martha C. 2003. *Upheavals of Thought: The Intelligence of Emotions.* 1. paperback ed. Cambridge: Cambridge Univ. Press.

Ofri, Danielle. 2014. *What Doctors Feel—How Emotions Affect the Practice of Medicine.* Boston: Beacon Press.

Parsons, Talcott. 1951. *The Social System.* Abingdon: Taylor & Francis.

Price, Sheri, Shelley Doucet, and Linda McGillis Hall. 2014. "The Historical Social Positioning of Nursing and Medicine: Implications for Career Choice, Early Socialization and Interprofessional Collaboration." *Journal of Interprofessional Care* 28 (2): 103–9. https://doi.org/10.3109/13561820.2013.867839.

Reed, Pamela G., and Cindy J. Rishel. 2015. "Epistemic Injustice and Nurse Moral Distress: Perspective for Policy Development." *Nursing Science Quarterly* 28 (3): 241–44. https://doi.org/10.1177/0894318415585634.

Silva, Joana Vilela Da, and Irene Carvalho. 2016. "Physicians Experiencing Intense Emotions While Seeing Their Patients: What Happens?" *The Permanente Journal* 20 (3): 15–229. https://doi.org/10.7812/TPP/15-229.

Sirriyeh, R., R. Lawton, P. Gardner, and G. Armitage. 2010. "Coping with Medical Error: A Systematic Review of Papers to Assess the Effects of Involvement in Medical Errors on Healthcare Professionals' Psychological Well-Being." *BMJ Quality & Safety* 19 (6): e43–e43. https://doi.org/10.1136/qshc.2009.035253.

Stockdale, Katie. 2024. "(Why) Do We Need a Theory of Affective Injustice?" *Philosophical Topics.* 51 1.

Waterman, Amy D., Jane Garbutt, Erik Hazel, William Claiborne Dunagan, Wendy Levinson, Victoria J. Fraser, and Thomas H. Gallagher. 2007. "The Emotional Impact of Medical Errors on Practicing Physicians in the United States and Canada." *The Joint Commission Journal on Quality and Patient Safety* 33 (8): 467–76. https://doi.org/10.1016/S1553-7250(07)33050-X.

Weilenmann, Sonja, Ulrich Schnyder, Brian Parkinson, Claudio Corda, Roland von Känel, and Monique C. Pfaltz. 2018. "Emotion Transfer, Emotion Regulation, and Empathy-Related Processes in Physician-Patient Interactions and Their Association With Physician Well-Being: A Theoretical Model." *Frontiers in Psychiatry* 9 (August): 389. https://doi.org/10.3389/fpsyt.2018.00389.

Wible, Pamela, MD | Physician | February 3, and 2014. 2014. "Are You a Victim of Patient Profiling?" KevinMD.Com. February 3, 2014. https://www.kevinmd.com/blog/2014/02/victim-patient-profiling.html.

Chapter 3
Patients and Affective Injustice

Abstract In this second chapter, I will pay attention to patient experiences of affective injustice and how this impacts care and their quality of life. This chapter builds upon the literature on epistemic injustice which has identified some of the affective harms that may come from patient's emotions and feelings not being taken seriously by their healthcare providers. In this chapter, I will first investigate and clarify the ways, means, temporalities, and spaces in which patients may suffer from affective harms in healthcare, notably when healthcare providers do not take them seriously or understand them. I will then focus on the various temporalities in which patients and their caregivers may be subject to affective injustice in the healthcare journey. I will also devote discussion to the affective harms that may come for persons who already face marginalization and exclusion and in which affective injustice may worsen their situations. This is important to fully understand the ramifications of affective injustice for all patients. At the end of the chapter, I will propose several possible remedy strategies, including some possibilities and reformulations from the epistemic injustice literature.

Keywords Patients · Affects · Emotions · Affective humility · Epistemic injustice

3.1 Introduction

The experience of healthcare—whether lived directly as a patient or by proxy as a caregiver—is an intensely emotional journey. It involves the surprise and horror of learning a diagnosis to the anxieties of becoming another that may no longer resemble our past selves, to the emotions that arise when deciding (non-ideal) treatment plans. The experience of illness emotionally affects not only the person, but also family members, colleagues, friends, and healthcare providers. How patients process, work through, and live with the new ways of being due to illness is both an individual and a collective emotional experience.

The affects lived and expressed by patients may be a valuable means of knowledge, if we can take them seriously and look at what they might be telling us. For instance,

B. Bogaert, *Affective Injustice in Healthcare*, The International Library of Bioethics 112, https://doi.org/10.1007/978-3-031-94378-2_3

while the anger felt at being diagnosed with a chronic illness may be hard to process in the immediate situation, in retrospect, it may help patients to better understand what this anger was telling them: they were scared, having an illness is unjust, they were anxious about how it will affect their lives, families, and career plans. Just as healthcare professionals have a hard time to accept their own affects, so too do patients, as it brings to the forefront vulnerabilities that are difficult to recognize and accept. I will argue in this chapter that affects can help a patient by showing them what is salient. Having the healthcare provider take them seriously also means that they can help provide the best care for their patient, as it allows them to understand the patient's illness experience in a holistic way.

There are some promising signs that healthcare providers are making efforts to recognize the importance of the patient's affects into care planning. The turn toward more empathetic care has notably helped healthcare providers to better understand that affects are an important part of the illness experience, including with narrative medicine models that seek to integrate the patient's life story—including their affective experiences—into care planning. Halpern (2001, p. 40) in her empathetic medicine model in particular gives prominence to these experiences and how they may improve care. She proposes that expressing empathy toward the patient can lead to better clinical decision-making, notably by showing curiosity and concern toward the patient. This model involves paying attention to what is salient from the other person's perspective (p. 74), including the affects that shape the patient's interests and beliefs. Halpern claims that without this, healthcare providers risk projecting our own values and ends onto patients (p. 114).

From this perspective, recognizing the patient's affects is one of the means of working toward a person-centered perspective, one that is focused on the person's values and ends. Halpern's perspective on affective attitudes is very close to what I am trying to achieve with this discussion on affective injustice. It involves taking seriously, hearing, and trying to understand the person's affects, what it may say about their overall lives and what this may imply in healthcare decision making. However, it is important to say at this point that it is not a question of imposing the sharing of affective attitudes from the patient to the healthcare provider. The patient should want to, see the interest, and be willing to share these affects. In return, it will be up to the healthcare provider to provide a facilitating environment in which persons may be able to express them, should they so wish. It is in this way that the patient will be able to exercise their affective capabilities, as the healthcare provider has provided the space in which they are welcomed. What we are interested in achieving through affective justice for patients therefore involves recognizing the patient as a person, in their strengths and vulnerabilities (which may be expressed through affects), and the possibility to recognize these affects as knowledge that may be important to care planning.

To be clear, I am advocating that affects have a place in shared decision making, in particular for chronic illness management, as they are forms of knowledge that may merit consideration in deciding treatment plans. This means that they have a place in the patient's decision-making with *all* of their healthcare providers, and not just with a psychologist or psychiatrist. At the current time, in the increased fragmentation of care, the task of taking seriously and working with the patient's affects is often "sub-contracted" to the psychologist or psychiatrist, who is referred to by the doctor or the nurse to deal with the emotional aspects of care. While the importance of the psychologist is not to be denied, I am advocating that the doctor and nurse—but also other "non-obvious" paramedical personnel such as the physical therapist—may also support the patient in their emotional journey. I advocate that a facilitating environment for expression of the person's affects is necessary in all healthcare encounters, and not just in the consultation with a psychologist or psychiatrist. If we can accept affects as a form of knowledge—one which may say something salient about the person's values, what is important for them, what they are afraid to lose—we can see that these forms of knowledge are important in all aspects of care delivery.

To move forward with this issue, in this chapter I will first explore what the epistemic and affective injustice literature can bring to understandings of the barriers that patients face in having their affects be heard and understood. I will then explore in more depth the temporalities in which affective harm may occur for the patient in the healthcare journey, including for marginalized groups. At the end of this chapter, I will also explore some remedial strategies, in particular for patients and their caregivers.

3.2 Patients Experiencing Epistemic and Affective Injustice

This section will look at the epistemic and affective harms patients may suffer from when they are not taken seriously as an affective being. At the time of writing, as affective injustice is a new field, I am not aware of any specific applications of how patients may be subject to affective injustice, other than developing work in psychiatry, which brings specific concerns (Lavallee and Gagné-Julien 2024). I advocate there will be a need going forward to be careful in these definitions, formulations, and applications to understand what affective injustice can add to the overall epistemic injustice debate, which has raised a number of these issues already. Therefore, in this section, I will first discuss the ways in which the epistemic injustice literature has conceptualized some affective harms, before describing its innovation by mobilizing several concepts specifically from the affective injustice literature.

In the first place, in their characterization of epistemic injustice, Carel and Kidd (2014) have already shown that epistemic injustice may occur when patients are considered "too emotional" to be taken seriously by their healthcare providers. Even though they do not use the words affective injustice, they describe the kind of affective harms that will interest us in this discussion. Carel and Kidd characterize situations in which a patient's testimony may be ignored or misunderstood because they express themselves in an emotional way. This may happen for example during the announcement of a diagnosis: as I will explore in the next section, healthcare providers often avoid discussion with the patient when they are perceived as "too emotional."

From this example, it seems that patient's affects have a hard time being accepted as valuable forms of knowledge by the healthcare provider. They are seen as an impediment to dialogue, as it may inhibit "rational" decision-making about a treatment. In this way, they suffer from the same ideas that prevent healthcare providers from expressing their emotions. However, I advocate that affects show what is salient for the patient, and they are an important means for patients to express what they feel, understand, or know about their illness and how it may affect their lives.

Given that affects are forms of knowledge (important for the patient, important for the care plan), the harms that come from not recognizing or understanding them are both epistemic and affective. They are a missed opportunity for the healthcare provider to a gain a valuable source of knowledge and for the patient to share it with someone who can understand and help them. This situation may bring what the affective injustice literature describes as *emotional disorientation* brought about by affect-related hermeneutical and testimonial injustice (Gallegos 2021). When patients' affects are not believed or understood, this means not only that their knowledge is ignored but also that their well-being is ignored. It may lead them to be further isolated in their illness experience and increase the already distressing affects they are feeling.

In addition to being silenced by healthcare providers, patients also make deliberate attempts to self-silence their affects in the consultation due to the same perceived ideas that healthcare providers may have about emotional expression. This makes their recognition by the healthcare provider even harder, even when there are good intentions on their part to provide a safe space. The reasons patients may decide to self-silence their affects are multiple: (1) social norms that encourage evaluations of affects as irrational and a correct or erroneous assessment of how their healthcare provider will receive them; (2) the person's beliefs that they need to "hold it together" in front of their healthcare provider (or in front of their family members or colleagues) in order not to appear weak; (3) as a protective measure to avoid discrimination and over-pathologization (see section on affective injustice for marginalized groups).

In the last chapter, I already explored the various emotional regulation strategies healthcare providers use. As we can see, these strategies are also used by patients, with similar harmful effects. This includes convincing oneself one is not feeling a particular emotion (cognitive reappraisal strategies) or trying actively to suppress one's facial expressions (emotional suppression strategies). The problem with this from the patient perspective is that affects are an inevitable part of the healthcare journey, as the decisions taken affect their lives, their mobility, their ability to have social relationships. When the patient tries to distract themselves or hide away from these affects, this is a missed moment for the healthcare provider to question what is behind them (what they value). As I have explored in Chapter 2, such types of emotional regulation strategies do not mean that they will disappear—but rather that the fear and anxiety patients experience may be channeled instead toward caregivers or other persons in their lives, and the relationships themselves may suffer because of it. In the context of chronic illness, where the support network is so important, this can have serious consequences for the person's and the caregiver's well-being.

From this discussion, we can see that patients experience affective injustice in at least two forms: (1) by not having their affects understood or taken seriously by a third party (healthcare provider, family, friends, colleagues); (2) by deciding to self-silence their affects, in the expectation that they will not be well received. Both instances of affective injustice lead to affective harms in terms of their well-being. It may lead to further isolation and possible rebound effects on outside relationships (in particular their support network). However, it can also have negative effects on their capability to participate in shared decision-making. It also means that there is valuable information missing from the healthcare provider's perspective—and that they cannot act in consequence to help them. To understand more specifically how this may occur, I will discuss how affective injustice may happen in specific temporalities of the healthcare journey in chronic illness.

3.3 Affective Injustice in Temporalities of the Healthcare Journey

Having elaborated in what ways patients may be subject to affective injustice, in this section I will explore the topic in greater depth by looking at the various ways patients may experience affective injustice during the healthcare journey. As the epistemic injustice literature has demonstrated, there will be likely be specific issues facing certain categories of patients. I will look at this situation later on in the chapter; however, in this section I have chosen to focus on common experiences of illness that may affect (nearly all) patients. There is good reason to believe that many patients share some common experiences of affective injustice in various temporalities of their healthcare journey, as they may face common emotional experiences from diagnosis to acceptance to living with their chronic conditions and how healthcare providers (and others) perceive and deal with them.

However, before starting the discussion, two points of clarification are necessary. First of all, I have chosen to focus on chronic rather than acute illness. While some of these issues will be similar, further work will need to be developed to study affective injustice specifically for acute illness. Secondly, I have chosen to use the word "patient" in this section in order to specify persons who solicit the help of the healthcare system to diagnose or treat their illness. However, I recognize that contact with providers is only a small part of the reality of living with a chronic illness. While for conceptual clarity, I have chosen to use the word "patient," I recognize that this term is not ideal as it does not give voice to the diversity of the person's experiences and how illness may affect them. I will return to this issue at the end of this section when I discuss the role of caregivers.

Having elaborated these limitations, I will now explore three moments in which affective injustice may occur for the patient in the chronic illness journey. The moments I will describe are the shock of the diagnosis; the acceptance phase; and living with chronic illness. As I will show, although the discussion is focused on patients, I will also pay attention to the lived experiences of support persons (healthcare providers, family and friends) who may live emotional experiences with them in this journey.

3.3.1 *The Shock of the Diagnosis*

The diagnosis of a chronic illness is the moment in which patient's affects are most visible to the healthcare provider, their family and friends, and patients themselves. The announcement of a diagnosis—even if suspected—often involves a shock. During this time, it is common for a patient not to hear—or rather not to remember—what they have been told. For those suffering from diagnostic delay—as is the case of rare diseases where there is considerable diagnostic wavering—there may be shock, but also a sense of relief: finally, the cause has been found and there may be something to do about it. Yet even for these patients, the announcement of the diagnosis may also be one in which it is difficult to appropriate all of the information coming at them. In these instances, the patient's affects (fear, anxiety) may be a barrier to further interaction and discussion.

Because of this, there is good reason for healthcare providers to be hesitant to relay all details of the illness to the person and start to engage in conversations about the healthcare plan in this temporality. They witness the grief, the shock, the fear and anxiety they induce in the patient by communicating the diagnosis and make an evaluation that it is not the appropriate time to engage in further discussions. However, the danger here (as already elaborated by Havi and Carel 2014) is if the patient is seen as "too emotional" and this leads their testimony not to be taken seriously. Given the intense affects experienced by the patient at this time, it may also be a key learning moment for the healthcare provider to gain valuable information about their patient.

What we can see from this type of situation is the tendency by professionals to apply the same criteria they have to their own healthcare practices (maintaining a certain amount of affective neutrality) and expect their patients to return in kind (by maintaining a certain objectivity in their discussions). It is perhaps for this reason that some may evaluate them as "not rational enough" to discuss their illness when affects are present. Let us explore why this kind of thinking may be problematic, even if we recognize that affects may be a blocking factor for some discussions and decision-making in this first temporality.

First of all, this first moment is vital to build a trusting relationship. Acknowledging what the patient feels may be the means to provide a safe emotional space for future interactions, even if shared decision making is not immediately possible. If the tendency would be at this point to refer the patient to a psychologist, this is a missed moment. Studies have shown that while referrals at this stage may have certain benefits, they also have a limited effect on the patient's psychological well-being, notably because of a mismatch between the patient's needs and provision of care (Dekker et al. 2020). In the first place, there remains considerable stigma by patients to address a psychologist: for many, it means that they have some problem that "needs to be fixed" because they are experiencing feelings about what is happening to them. If patients believe their affects are something to be corrected, this may actually make the situation worse. This can cause affective harms to the person's well-being, but it may also lead to epistemic harms in the consultation, as the patient may decide to self-silence them and other information in future encounters.

There is also the question of access, as psychological consultations are not reimbursed in many countries, making it a care service that is difficult to access, in particular for vulnerable groups. If these persons neither receive emotional support from their healthcare provider nor from a psychologist, they may be left alone with troubling emotions. Delegating this task to a psychologist also means that healthcare providers (specialists, but also primary healthcare providers, or regular healthcare providers such as physical therapists) are missing out on a valuable moment of knowledge and experience that may be important for shared decision making. They may also be missing a moment in which a collaborative relationship between the patient and the healthcare provider may be built, one based on trust.[1] Therefore, while the psychologist can work in a more long-term temporality with the patient and their emotional needs as they come to terms with their illness, it is also important for other healthcare providers to be able to welcome these affects in the consultation. As Dekker et al. (2020) argue, patients experiencing adaptive affects (such as the normal anger that may come from learning of a diagnosis) may welcome support from relatives, friends, and healthcare providers to talk about these affects, but feel uncomfortable or find it strange to seek care from a psychologist.

From this perspective, an affective resource can be created for these patients in the (already) existing healthcare relationship. Asking the patient, "do you feel ready to discuss these issues?" "What can I do to help you?" "Can we make a further appointment to discuss how you can feel?" may be enough to build this space. It can also help the patient recognize that it common, normal, and even important to recognize the affects they are experiencing, particularly for those patients who try to put on "a brave face" in front of their healthcare provider. Thus, if the patient knows that their affects will be taken seriously by their healthcare provider, they can also start to work with them and establish a relationship of trust for future interactions. In addition, the danger of not recognizing these affects in this diagnostic encounter is that for some, it is indeed the fortuitous moment for them to start talking about their care, at least to be able to relate what has been said to their family and friends. Not recognizing them or putting them aside by delegating them to another professional can therefore contribute to their emotional disorientation in a scary time.

Before moving on to the next temporality of the illness journey, it will important to take a moment to acknowledge how these diagnosis moments affect healthcare providers themselves. Learning how to announce a diagnosis is an important part of medical training, and it causes considerable anxiety and stress for students. While they are trained to maintain "detached concern" (Fox 1998) in the face of such announcements, seeing how such information affects patients is a difficult time for

[1] One concrete example of the importance of paying attention to the patient's emotions in a consultation is adherence. By not addressing the patient's fears of the medication—and the inevitable side effects it may bring—healthcare providers are unlikely to clarify the situation and the worries of the patient, as well as miss a valuable moment to communicate what to do should these fears be realized. The psychologist here is not necessarily the best person to deal with the patient's affects—here it is for the doctor to take these seriously and try to communicate with the patient to help them overcome their fears, in particular by being honest and forthright with the information they have and by putting in place safeguards in case of side effects of medication.

any healthcare professional, no matter the number of years of experience. In these temporalities, emotional regulation strategies may be used by the healthcare provider to give full attention, value, and priority to the patient. They may be a way to reassure the patient by not having them see the fear or distress they feel mirrored in their healthcare provider, which may (for some patients) lead them to believe that their healthcare provider is not equipped to help them. In this temporality, there is a value in the emotional regulation strategies by the healthcare provider, as they are turned toward the best interests of their patients. However, engaging in such strategies does little to pay attention to the healthcare provider's needs. Here, the institution can go a long way toward helping the professional (and thereby the patient) by providing safe spaces outside of the consultation to discuss these difficult moments. After all, taking care of patients and their emotional needs also means taking care of healthcare provider's needs. I will explore some means to achieve this in Part II of this book.

Having explored the affects lived by the healthcare provider and the patient in the announcement of a diagnosis, let us now turn to the path toward acceptance or at least acknowledgement of an illness.

3.3.2 The Path to Acceptance

The phase to acceptance of one's chronic illness is a rocky, unpredictable, and emotional one—and in some persons, acceptance may never fully happen. This may be hard for others to understand. Acceptance is not a linear path, one in which the person follows a straight line and will never return back to the start; it is often necessary to move back and forth and find one's own path to acceptance—and understanding what this acceptance implies. During this uncertain process, patients experience a range of well-known affects, from denial and anger to depression. However, even once they have accepted an illness as a reality, this does not mean that these and other affects will not surface.

In this stage, a common affect is denial or disbelief, such as thinking, "this must be a mistake" or trying to find ways to avoid thinking about it. Another common one is anger: directed at oneself, at the world, or even those closest to the person. It is also common to experience depression in these situations, which involves feelings of sadness, hopelessness or helplessness due to an uncertain—and often difficult—future. What is particularly problematic (for the person themselves but also for those trying to support them) about these affects is that while there is an often a common pathway which can be understood, each person may experience it differently. It may also be the case that some affects may seem to creep up at strange times, in bizarre circumstances, or in reaction to those things which others deem inappropriate, such as in a discussion of treatment options with their healthcare provider.

It is during these moments in which patients may be particularly vulnerable to affective injustice by their healthcare provider, in particular when the provider has judged that the patient should be ready to engage in shared decision making in a "calm," "rational" or "objective" manner. While some may understand and empathize with the anger or sadness the patient may feel about their diagnosis, there may be less acceptance of affects in this temporality, notably because it involves making difficult choices, such as weighing the benefits and risks of a given intervention. For healthcare providers, affects may seem to be inappropriate or a sign of inadaptation or non-acceptance. From the doctor's perspective, after all, there is a treatment. After all, there are clinical trials to attempt. Why then, do they still have such negative feelings? Isn't this a sign of giving up? Shouldn't we encourage them to have hope?

Most of us have lived through this kind of situation—either as a patient, a friend, or a family member. We have lived the frustration of trying to support a person that we care about, and it is particularly trying to support them when we believe that their anger or frustration should be gone. In these situations, whether as family members or healthcare providers, we may be unwittingly perpetrators of affective injustice. First of all, by making an evaluation on the appropriateness of an emotion, we have decided—in place of—the patient that they are ready to leave their fear and anger behind. Because of this, support persons can (at times) also (unintentionally) be perpetuators of affective injustice if they are not willing to hear and understand the person's affects and what it may be saying. The risk is that in these situations, we also may be blocking the opportunity for further dialogue and support, leaving them isolated (and even more) distressed, even when our evaluation of the affect may be justifiable or understandable from (our) perspective as a support person.

While this kind of situation can bring complexities to relationships between the person facing chronic illness and their support network, it takes specific forms in the consultation. Here it is worth asking if it is up to the healthcare provider to decide what is appropriate for the patient to be experiencing, or to be there as a witness, to hear and understand. Furthermore, given the difficulties of dealing with these affects in daily life, caregivers may be less tolerant to understanding and recognizing these affects because of the illness burden on their lives. When this happens, the healthcare provider may be an important affective resource for patient, as they are less burdened by the daily life of the illness and may be more willing to listen and understand about the challenges the patient is facing.

In addition, during this temporality, there is often an expectation that patients must move beyond negative affects—fear, anger—and to work toward more positive ones, and in particular to cultivate hope. In chronic illnesses, patients are encouraged to "never give up hope" that they will eventually feel better and get better, even in the case of non-treatable illnesses, as science is "always developing" and "scientific progress being made." Developing hope is even considered an important stage in the grief cycle and in the path to acceptance. Not moving beyond one's fear or anger is therefore seen as never fully accepting one's diagnosis or falling into a sort of fatality that may make the illness even worse. Forcing certain affects upon patients (who are not ready and may never be), in particular by forcing more "positive-sounding" affects on them, may be misplaced at best, and harmful for the patient themselves and their understanding of their illness. It may lead to further emotional disorientation. In addition, when their emotional experiences do not fit with the dominant hermeneutical framework, this will likely isolate them even further. To use the vocabulary from the affective injustice framework, this may lead to hermeneutically related affective injustice, which may make them doubt what they feel as it is does not fit what is expected of them.

What we can see from this discussion is that the responsibility is on us all—all support persons, whether healthcare providers, family members or friends—to gain a critical awareness that we may be committing affective injustice toward these persons—even with the best intentions. It is our responsibility to recognize and give the safe space for affects by letting them know we hear and understand them. And as we have seen, forcing certain affects such as hope upon to patients may also be a particularly harmful form of affective injustice.

3.3.3 Living with a Chronic Illness

I now come to the last stage of the healthcare journey, which I define as living with a chronic illness. The illness has been diagnosed; it has been accepted. Now what? Learning to live with a chronic illness is both a time of intense affects—but it is again one in which affective injustice may occur both inside and outside of healthcare. It is expected that once a treatment is in place, once patients "feel better," once they have accepted their illness, that patients will no longer show—or even experience—the intense affects that were tolerated in the past. Yet, the period of living with a chronic illness is still an emotional rollercoaster and one in which affective injustice continues to occur, both in the consultation and outside of it. In this section, I will first discuss the more medical aspects of the treatment plan before turning to affective injustice in the person's overall life.

To return to the question of hope, as I have already started to discuss, patients have been socialized in many countries to take faith in hope, particularly from family and friends who encourage "positive thinking." Optimism and hope are encouraged both by caregivers and wider society, even beyond the power of medical progress, as there is some evidence that hope may lead to greater physical and psychological health in some chronic conditions (Schiavon et al. 2017). For instance, having hope may encourage the person to engage in physical exercise, which may help lessen the severity of cardiovascular and other chronic conditions. It may also encourage patients to learn about their illness or to participate in clinical trials and other forms of experimental treatment, research which may help themselves (potentially) but will in particular help future patients. However, there is a difference between fostering realistic and non-realistic hope, and this discussion is closely related to an ethics of transparency with the patient. On the one hand, the danger of unrealistic hope is that it may be inappropriate to the situation at hand and lead to false illusions which will eventually be deceived, leading to greater emotional distress. When the patient experiences this feeling of deception (for instance a failure of a clinical trial), they may feel let down and it may even be a harder blow than not having this hope in the first place. In turn, even when the doctor is transparent about the fact that current treatments are unlikely to help them, patients have been educated and encouraged to believe by society—and their close friends, relatives, and colleagues—that a treatment or a cure will eventually be found. The tightrope between encouraging hope and being fully transparent toward the patient is difficult to manage, particularly because healthcare providers themselves often entertain the same hope of one day finding a cure.

In these situations, affective injustice may occur for the patient when they are not taken seriously by the healthcare provider for entertaining unrealistic hopes in what the healthcare provider perceives as a "hopeless" situation. Here healthcare providers are making a judgement of the patient and the inappropriateness of their emotion, causing them to dismiss it or not to give adequate attention to it. This kind of situation may be particularly frustrating to a healthcare provider who has done their best to be forthright with their patient. However, even if they believe this hope is unrealistic, healthcare providers may also decide (in a condescending way) to encourage it because they believe that hope is good for the patient's health. Placating the patient by letting them entertain unrealistic hope does little to help the patient to deal with the ramifications of their illness and does not take them seriously as affective and epistemic beings ("we can let them hope, after all it's good for their health!").

We can find a path forward from this difficulty if we return to the idea of affects as forms of knowledge. From this perspective, the role of healthcare providers will be to help the patient toward greater affective capabilities. By this term, I am suggesting that it is up to the healthcare provider to provide a facilitating environment in which an open expression of their fears, anxieties, hope, and worries may be given space and recognition. In the case of hope, it can help the healthcare provider to better understand the source of this hope and how it relates to the person's values and beliefs, including spiritual beliefs outside of medical understandings. Open discussions of hope may therefore be harnessed to help the healthcare provider to better understand the patient's motivations and values. Such a stance may also help the healthcare provider help the patient toward a more realistic hope, or what could be termed "healthy coping" (Turner 2000). To foster healthy coping, it is important for the healthcare provider to provide such a facilitating space, one in which the patient can exercise their affective capabilities, in particular when they lack the affective resources outside of the healthcare provider relationship, notably because they are encouraged by family and friends to stay "positive."

To conclude this discussion, we can see here that when "things can return to normal" or at least to a "new normal," affects are still present and important. They say something salient about the illness experience which is not to be ignored. They relate to pressure on persons who are attempting to juggle both a chronic illness and the realities of illness on their lives, as well as social expectations of them. Being able to express these affects in the consultation in this way helps work toward a more person-centered perspective by better understanding what motivations, values, and beliefs are important to the patient. Treating them as a form of knowledge and as important to shared decision-making helps work toward this goal.

3.3.4 The Role of Caregivers

Before concluding this section, it will be important to briefly explore in what means and temporalities caregivers may experience affective injustice in the illness journey. I use the general term caregivers to refer to those persons involved in the patient's care, whether as a spouse, friend, or family member. These experiences can be diverse and changing; however, I am most interested here in those supporting the person regularly and in a long-term temporality, notably because of the high emotional toll of caregiving.

In the first place, it is to be acknowledged that caregiving involves significant changes in the relationship between the person being cared for and the person caring. Close friends or family relationships change dramatically when caregiving roles are added. These new roles are difficult to manage both for the person who has the chronic illness (and finds themselves in a situation of dependance toward their caregiver), but also for the caregiver themselves, who often find themselves wanting to support their loved one, but ill at ease with assuming caregiving roles. These changes influence how the caregiver interacts with their loved one; it also influences how the person with the chronic illness sees and treats them. Situations of increased dependance are difficult to deal with for both persons and strong affects—such as anger—inevitably occur.

However, while the affects of the patient may be accepted (in some temporalities, as we have seen) as a natural part of coming to terms with a chronic illness, the affective needs of caregivers are often neglected by others, but also by themselves. Caring for a loved one with a chronic illness may help loved ones come together; however, it can also be overwhelming, both physically and emotionally. Caregiving can result in chronic stress and depression (Schulz and Sherwood 2008) and caregivers may experience negative affects such as anger (at their partner, at the illness, at their spiritual beliefs), depression and regret, deep sadness and grief about the loss of the life they had envisaged with them and their loved ones (Sullivan and Miller 2015).

In these situations, caregivers may believe they have to prioritize their loved one and what they are experiencing. It is harder to accept (and recognize) how this experience may be taking an emotional toll on them. It is even harder to openly express their affects to others, especially their loved ones. It is for this reason that caregiving has been called "emotional work." It involves both the self-management of affects and adopting and embracing a positive attitude when providing care and support. Some of these strategies include putting the needs of the person with the illness before the needs of the caregiver, protecting them or avoiding conflict (Herron et al. 2019). It also includes emotional regulation strategies, such as separating their "true" feelings of anger and frustration and what is acceptable to express (Halevi et al. 2022). These strategies involve deflecting attention away from their emotional needs in favor of the patient/loved one. These are similar to the emotional regulation strategies employed by the healthcare provider, who tries to suppress their affects to work in the best interests of their patient. However, it leads to similar affective harms in terms of well-being and increased risk of burn-out.

From this short discussion, we can see here the importance of developing affective resources, spaces, and opportunities for caregivers. While there is an increasing awareness of these problems, there remains few resources at the current time in healthcare institutions specifically for caregivers. This is particularly difficult as caregivers themselves may see their own affects as illegitimate and try to suppress them. Clearly strategies need to be found specifically for caregivers if we wish to achieve affective justice for them.

3.4 Affective Injustice: Certain Groups May Be More Vulnerable

Our discussion thus far has taken a wide approach to better understand common patient and caregiver experiences of affective injustice. However, it will be important to take a moment to better understand the affective harms that may occur for certain groups due to wider factors of exclusion. As the epistemic (and now affective) literature has shown, the criteria of a "patient" may already be a devaluing factor; however, the interaction of the social status of a patient with other factors of marginalization may further increase epistemic and affective harms for certain groups. As I will show, in some cases, due to the person's social status, their emotions may even be regarded as symptoms of psychiatric disturbance, or what Pismenny et al. (2024) calls emotional pathologizing. This encourages self-silencing to avoid further discrimination.

By marginalized groups, I refer to those persons who are excluded from mainstream social, economic, educational and/or cultural life. Examples include (but are not limited to) groups excluded due to race/ethnicity, gender identity, age, language, or immigration status (Sevelius et al. 2020). Marginalization in healthcare means that these groups suffer from lower access, less quality care, and more conflictual relationships with their providers. They are subject to bias and stereotyping, which has an impact on how they are treated and understood by healthcare providers. Unfortunately, it has been demonstrated that healthcare providers manifest bias to a similar degree as the general population, making them treat patients differently in function of their race/ethnicity, gender, socio-economic status, mental illness, weight, having AIDS, disability, and social circumstances, among other factors (FitzGerald and Hirst 2017). This bias leads to situations of affective injustice. This may be the case for instance when cultural differences and beliefs do not correspond with healthcare provider's expectations of how they should act, making the patient's affects seem inappropriate in the situation and resulting in them not being believed or understood.

Getting a better handle on these additional vulnerabilities will be important to address the full extent of affective injustice in healthcare. While the discussion thus far has focused on how affective harms may hurt the individual person's well-being, these harms also have a wider impact on society, as it means that non-dominant experiences are missing from our collective understanding and that healthcare exasperates these forms of exclusion. In this respect, affective injustice is one of the ways social exclusion happens in practice.

Because this is a wide debate, I will focus here on two issues as a way of illustration of the extent to which affective injustice happens for certain groups. The first is how gender bias can lead to affective harms for both men and women. The second issue is how an intersection of factors—gender but also race/ethnicity—may make certain persons more suspectable to affective harms by healthcare providers.

3.4.1 Affective Injustice and Gender

As we have seen, there is a persistent idea that healthcare providers cannot engage with patients when they are too "emotional." However, there are some patients that may be disproportionally affected by this, particularly when they are seen as overly emotional as a social group. In this section, I will argue that the persistent idea that women are more "emotional" and that men are more "stoic" brings affective harms for both genders.

To start our discussion, it is well-documented that gender bias continues with ongoing perceptions of women as more "emotional" than men. Despite decades of advocacy for gender equality, there remains an undercurrent that supposes women are less rational than men (as they are supposedly more "emotional"). Given that healthcare providers are part of societies which hold these explicit and implicit beliefs, it is to be expected that these biases will have ramifications on how healthcare providers see and treat women's affects. Differential healthcare treatment due to gender differences and in particular the perception that women are emotional (and thus irrational) was already demonstrated from Antiquity. Influential thinkers such as Aristotle did not even recognize a difference between male and female bodies, or rather they hypothesized that the female body was underdeveloped compared to men. This led to ideas that there were no "women's diseases" and little research on gender and sex differences. While Hippocrates did recognize the possibility of conditions which could specifically affect women, he added fuel to the fire by claiming that womanly conditions were based on affects. Even more, he believed that affects overwhelmed women and lead to them becoming "hysterical." (Lippi et al. 2020).

These ideas of women's health being linked to excessive affects have prevailed to a certain extent even today. Women are more likely than men to be diagnosed with a psychological (rather than a physiological) cause for their condition, with serious repercussions for healthcare outcomes, from diagnosis to pain management. Numerous studies have shown that women are less likely than men to receive more advanced diagnostic and therapeutic interventions. This is the case for a large variety of illnesses, including coronary artery disease, Parkinson's disease, irritable bowel syndrome, neck pain, knee joint arthrosis, and tuberculosis (Hamberg 2008). A particular telling example is the gender bias in pain management. Hoffman and

Tarzian (2001) have shown that women tend to be more diagnosed with psychosomatic illnesses and given sedatives rather than stronger pain relief compared to men. Women are also less likely to be believed in illnesses for which there is currently a lacune in medical understanding, such as fibromyalgia (Manne 2021, p. 81). This also has an impact on medical research, as data has been historically gathered almost exclusively from white male participants, leading to biases in treatment of illnesses.

From these examples, it is evident that gender bias has a significant impact on the unequal treatment of women in healthcare (let alone transgender or intersex persons). While the perception and treatment of affects are not the only factor to understand these disparities, they are important to the debate for two reasons: (1) women may hesitate to show their affects to their healthcare provider due to a perception of gender bias by their healthcare provider; (2) healthcare providers may judge affects expressed by a woman unfairly due to prejudice. The woman will be in both cases harmed as an affective being. However, as we can see in the example of pain, such perceptions should not only be understood in terms of well-being. As we can see, it also leads them to receive less quality care, which will (inevitably) lead them to have less quality patient-provider communication and trust in the healthcare system. Furthermore, ignoring or misinterpreting a woman's affects may also have ramifications on healthcare outcomes, leading the healthcare provider to misinterpret or miss out on important diagnostic clues, particularly when they decide these affects are psychological rather than physiological because they are coming from a woman.

However, it should also be recognized that the idea of women as more emotional than men also has repercussions on the affective capabilities of men. A systemic review has shown that healthcare providers perceive men as "stoic" and more "tolerating of or denying pain" (Samulowitz et al. 2018). Given affective norms, men may engage in emotional suppression strategies in front of the healthcare provider to conform to these social expectations. They may also have a harder time expressing these affects in other contexts, as their parents, the educational system, and wider society has discouraged development and expression of their affective capabilities. This may lead to affective harms for the man, who may experience vulnerability, distress, and suffering, but is unable to discuss (or even recognize) what he is experiencing. Given gender bias however, the man who is willing to express their affects may also be more likely to be taken seriously by the healthcare provider, as it may strike the healthcare provider as an unusual or an unusually important sign that the man has expressed these affects. This happens for example with gender differences in treatment for pain. As we can see, social expectations can go both ways, either in favor of or to the detriment of the person's emotional needs. These discussions have however shown the danger of having affects not taken seriously or misunderstood by the healthcare provider due to gender bias and how this causes serious affective harms to both men and women.

3.4.2 Affective Injustice, Gender, and Racism

We have seen the difficulties women have to have their affects be heard and understood in healthcare. However, among women, some are even more vulnerable to affective injustice due to other factors of exclusion, in particular due to the intersection of gender bias and racism. To illustrate this problem, I will take the example of African American women's perceptions and use of mental healthcare. Due to discrimination and other factors of exclusion in society, these women are disproportionally affected by feelings of sadness, depression, and psychological distress. This should make them a priority group for mental health care services. However, they are also the least likely to access these services. The reasons for these are multiple and related to historical mistrust of the healthcare system, ideas on mental health, expectations of women, but also the realities of the unequal care provided to them. This example will therefore show us the emotional suppression strategies these women engage in both due to social expectations and mistrust in the medical system, and the resulting harms that this may cause them.

In the first place, these women are reluctant to access mental health services due to concerns about their healthcare provider's attitudes toward them. They experience dissatisfaction with a healthcare system in which discrimination and bias toward them is recurrent (Copeland and Synder 2011). As Holden et al. (2014) have shown, this group is underdiagnosed and undertreated for psychological disorders but correspondingly over diagnosed and overtreated for psychotic disorders. In these situations, the way that the woman's affects are understood and treated is linked to both gender and racial biases. According to Woods-Giscombe's qualitative research (2016), women from this community are concerned with what would happen to them should they fully express their emotional distress. According to one participant from their research, "I don't tell people because they start to think you're kinda going crazy!... you kind of got to watch what you tell people." These women believe (and research correlates with this view), that it is better if women suppress their affects to avoid being emotionally pathologized (Pismenny 2024) by their healthcare provider. This is a rational coping strategy and it is a reality of how healthcare is delivered.

Bias toward this group is in particular caused by the "angry Black woman" stereotype, which falsely sees these women as aggressive, overbearing, or mean. The expression of affects by a woman from this social group—such as anger—may be easily misunderstood, particularly by the mental health professional who is unaware of their own stereotypes and how it may affect their attitudes toward patients. Research has shown that clinicians who do not pay sufficient attention to their own stereotypes may misinterpret data and symptoms, leading to false diagnoses, inadequate treatments and overall poor-quality care (Ashley 2014). This leads some women to try to suppress their anger in the consultation, and the affective harms that result include being left with feelings of helplessness, hopelessness, and even self-hate.

Other cultural factors also play an important role in the emotional suppression of this community, a phenomenon that Woods-Giscombe et al. (2016) have described as the "superwoman" role. The superwoman role involves suppressing one's affects but also resisting feelings of vulnerability and dependence. It forms an important part of the social expectations influencing how African American women act in their community and toward their healthcare providers. The problem is that healthcare providers are not necessarily aware of this and they may not take it into consideration.

As we can see, emotional suppression strategies are a protection strategy to keep from being misjudged or mistreated due to prejudice. They are also internalized due to cultural factors. However, just as in other suppression strategies we have seen thus far, they can have severe consequences for the person. According to Woods-Giscombe (2016), many women interviewed had experienced a breakdown because they had to suppress their affects. In these situations, there is both an affective harm in terms of the person's well-being, as well as an epistemic one, as it means that their knowledge is not sufficiently heard or understood in the treatment plan, with corresponding negative effects on healthcare delivery.

According to Woods-Giscombe's research, another factor in the community's mistrust of their healthcare providers and decision to engage in emotional suppression strategies (or not to seek help at all) was based on a perceived lack of culturally sensitive providers. Their research showed that these women believed their healthcare providers were lacking the understanding of their life experiences and therefore the compassion and capability to work with them, particularly in accordance with their values (including spiritual values). From this information, we can understand that the decision to self-silence affects was also a way of protecting their own affective knowledge from those who may not necessarily believe or understand them. The danger of this strategy will be that the person may not necessarily have sufficient affective spaces and resources in their community, particularly because of the "superwoman" role that is expected of them. In terms of the healthcare plan, it also means that affects important to mental healthcare may not be taken into consideration. Clearly, healthcare providers and healthcare institutions need to be humbler about different experiences of illness and try their best to better hear and understand the affective needs of this community. Working toward affective justice for these patients also includes making more systematic changes, notably by encouraging more diversity in medical students to enable these women to find providers they can trust.

3.5 Ways Forward: Safe Spaces and Persons for Affective Justice

As the proceeding sections have shown, affective injustice is a reality for many patients and their support persons and is in particular a problem for marginalized groups. While many of the remedial strategies to be detailed in Part II of this book may also benefit patients and their families, such as paying attention to affects in hospital architecture and design or creating safe spaces for narration, there is a need to specifically focus on what we can do about affective injustice for patients in particular. It is first of all to be acknowledged there are already a certain number of remedial strategies that exist from the epistemic injustice literature to help work toward affective injustice. In this section, I will discuss a few of these affective tools and resources even though they have not been called as such at present. Part II of this book will explore other strategies that may help patients to have greater affective resources, opportunities, freedoms, and recognition.

3.5.1 *Existing Tools and Concepts from the Epistemic Injustice Literature*

While the field of epistemic injustice has largely remained a theoretical and not empirical field (although this is changing), philosophers and social scientists interested in the concept have actively sought to provide remedial strategies. This is because the concept, while providing theoretical innovation, is also one focused on the non-ideal world and its problems. As it has helped to identify the problems, processes, and persons involved in epistemic injustice, it has (almost naturally) led to proposals to work toward an ideal of epistemic justice. In the application of the concept of epistemic injustice to healthcare, philosophers have suggested in particular two promising concepts: the idea of cultivating greater epistemic humility among healthcare providers (Wardrope 2015) and developing new methods to facilitate understanding of and expression of patient experiences through phenomenological methods (Carel 2012). I propose that these tools may be adapted to work toward affective justice for patients. Both of them involve an engagement at an individual and institutional level.

3.5.1.1 Affective Humility

Alistair Wardrope, an academic junior doctor in the United Kingdom, is well-placed to understand that healthcare providers may need to be humbler in their relationships with patients. He has strongly advocated that in order to combat epistemic injustice in healthcare, we will need to recognize the limits of our knowledge and actively seek out new information. His proposition of *epistemic humility* to work toward greater epistemic injustice in healthcare (Wardrope 2015) is defined as, "an attitude of awareness on the limitations of one's own epistemic capacities, and an active disposition to seek sources outside one's self to overcome these shortcomings." The idea of developing greater humility is starting to be discussed by healthcare providers and institutions; however, it remains a subject that is difficult to put into practice, in particular because doctors (in particular) are often trained and taught to be the epistemic authorities in healthcare. Developing one's humility, however, may be a key resource to help better understand and work with patients' affects, as I will discuss.

In terms of our discussion of affective justice for patients, the idea of epistemic humility is certainly relevant, as it means that healthcare providers will practice an attitude of modesty in their own knowledge and make an active search for other knowledge—and I can add here patient's knowledge, which includes the affects they experience. If I modify this term to encompass *affective humility*, I will be able to focus specifically on the attitude of modesty toward the patient's affective experiences. I therefore define affective humility as *the awareness that we do not necessarily experience and express the world in the same ways and an active disposition to look for and understand the other's affective experiences.*

In the context of healthcare, cultivating affective humility means acknowledging our incapacities to live through the other and fully understand their experiences. It means encouraging a disposition to actively find ways to better understand and listen to the patient. In the context of the healthcare provider-patient relationship, this can mean paying active attention to and trying to understand the affective experiences of the patient (or caregiver). This includes but goes beyond merely taking a better clinical history and recognizing that the other is incommensurate—and therefore not fully accessible to our worldview. Recognizing someone as an end means recognizing that they have separate experiences and ways to deal with these experiences. Affective humility therefore means paying attention to and supporting the patient to express their individual ways of being and understanding of the world and an engagement by the healthcare provider to try their best to hear and understand them.

In his epistemic humility concept, Wardrope stresses the engagement of the healthcare provider to work with different perspectives that go beyond their own legitimate (expert) domains, but also their collective responsibilities to work toward shaping collective conceptual resources and building a safe space to admit one's limits. In this perspective, working toward affective humility means not only bringing a critical self-awareness of one's limits and a disposition/desire to seek out other perspectives, but also an engagement toward the patient. It means helping the patient find appropriate resources and support persons to help them when they are distressed or in need of further emotional support. This person can be the psychologist, but it can also be other support persons and resources (patient associations, other healthcare providers) or as I have been arguing, it could simply mean providing a few more moments with the patient to express, honor, and help to understand what they are feeling.

In the context of those suffering from even greater barriers to affective injustice (as we saw for example in the case of African American women accessing mental health services), there is an even more pressing need to develop this kind of affective humility. This needs to happen at all levels, from including persons from more diverse cultural groups into medical training (to help these communities find persons with whom they feel more comfortable and who can better understand their affective needs), to having the whole community take a collective step back and realize that they will never fully understand the experiences of all their patients. It also means welcoming new perspectives into care planning, such as spiritual care (more on this in Part II) to incorporate the affective needs of a more diverse group with different values and needs.

If healthcare providers—but also healthcare institutions—can recognize the need for both epistemic and affective humility, and if they develop programs to help meet this need, we will be helping not only patients, but also healthcare providers, families, hospital administrators and managers. While I will discuss in more depth some of the possibilities in Part II of this book, for the moment, let us turn to a tool that can help all healthcare actors to better understand, develop, and express the affects they experience in care.

3.5.1.2 Carel's Phenomenological Toolkit

Philosopher Havi Carel, a leading philosopher of epistemic injustice in healthcare, has also provided a fruitful affective resource for patients and their families. Her proposition is a phenomenological toolkit (Carel 2012). Carel says that her model is not pre-defined, but rather provides a flexible and individualized tool which patients can use to develop understanding of their illness. The method is a means of providing a safe space for emotional (and other forms of) expression. Carel suggests for example the use of images and sounds to bring forth unnamed affects and experiences, or to ask patients to create art or choose a song to describe how they feel. Such methods may both help bring awareness to (and help the person see) the value of their emotional experiences in understanding and living with their illnesses, thus helping to develop their affective capabilities. They may also serve as an alternative forum to express illness experiences to others and make them understood. For instance, she suggests that listening to a song that helps explain someone's sadness or anger is not only more understandable—but also more memorable—and can help to better understand what they feel. Even more, when this is done in a group setting, the group can actively encourage these emotional expressions and in this way provide a safe space for its expression, helping patients to come out of affective silences.

The second step of the toolkit, which Carel calls "thematizing" is particularly relevant to our discussion of affects. It is a way of making particular aspects of a phenomenon explicit by paying attention to cognitive, emotional, moral or aesthetic experiences (Carel 2012, p. 108). She says, "the patient may thematize her illness emotively, while a health professional will thematize it cognitively." What is striking about this characterization is that she places the emotional experience alongside other aspects and means of knowledge, rather than in a hierarchy or separately. In this way, the cognitive experience of the healthcare provider is next to, not less than, the emotional experience of the patient.

However, while Carel suggests that her toolkit is a resource for patients, it may also be an affective resource for healthcare providers and caregivers to better understand and express their affective experiences in healthcare. These toolkits may help caregivers to communicate what they are feeling to their loved ones or at least to give a safe space to acknowledge what they are feeling to themselves. For healthcare providers, it may provide a needed outlet and safe space to explore with themselves and others the emotional impact of their work. What is also interesting about her toolkit is that it can both be adapted to in-hospital use (the patient in the hospital bed or in therapeutic patient education) but also take place in safe spaces outside of the hospital, such as in patient and caregiver support groups (more on this subject later). However, for it to have an impact on the healthcare provider's affective needs, it will likely need to be implemented directly in the hospital to become a realistic resource for them. Welcoming such programs into the hospital space for all—patients, healthcare providers, caregivers—is also a means to provide a greater affective recognition of these experiences in healthcare and their importance to quality care and well-being. While this may seem difficult in a time of increasing workload and decreasing resources, given the importance of providing such resources in order to foster well-being and as a stop-guard against burn-out (in particular for healthcare providers), it is worth considering how they may be integrated into daily hospital work. From my perspective, what is interesting about this resource is that it does not necessarily take much time: it may be done during a break or during a team meeting. It can also be integrated into trainings of healthcare professionals, providing a needed break and other way of improving care in trainings that are geared toward more technical aspects or specialized knowledge of care. Given resource shortages, this is likely the most feasible means of uptake and support for such affective resources in the hospital.

3.6 Conclusion

In this chapter, I have begun to conceptualize what affective injustice may look like from the patient's and caregiver's perspective. I expect that further work will be done in the coming years to characterize various forms of affective injustice according to the type of illness and their special needs and concerns. I have chosen here to focus on common experiences of chronic illness to bring attention to the problems patients may face across specialties, as well as bring a focus on the particular difficulties facing marginalized groups. I have also started to look at some of the affective tools, resources, and opportunities that may help affects to be better recognized and understood both inside and outside of healthcare. It will now be my job to explore some promising affective resources, opportunities, freedoms, and forms of recognition to work toward greater affective justice in healthcare. This is the ambition of the second part of this book.

References

Ashley, Wendy. 2014. "The Angry Black Woman: The Impact of Pejorative Stereotypes on Psychotherapy with Black Women." *Social Work in Public Health* 29 (1): 27–34. https://doi.org/10.1080/19371918.2011.619449

Carel, H. 2012. "Phenomenology as a Resource for Patients." *Journal of Medicine and Philosophy* 37 (2): 96–113. https://doi.org/10.1093/jmp/jhs008

Carel, Havi, and Ian James Kidd. 2014. "Epistemic Injustice in Healthcare: A Philosophical Analysis." *Medicine, Health Care and Philosophy* 17 (4): 529–540. https://doi.org/10.1007/s11019-014-9560-2

Copeland, Valire Carr, and Kimberly Snyder. 2011. "Barriers to Mental Health Treatment Services for Low-Income African American Women Whose Children Receive Behavioral Health Services: An Ethnographic Investigation." *Social Work in Public Health* 26 (1): 78–95. https://doi.org/10.1080/10911350903341036

Dekker, Joost, Jeanet Karchoud, Annemarie M J Braamse, Hilde Buiting, Inge R H M Konings, Myra E Van Linde, Claudia S E W Schuurhuizen, Mirjam A G Sprangers, Aartjan T F Beekman, and Henk M W Verheul. 2020. "Clinical Management of Emotions in Patients with Cancer: Introducing the Approach 'Emotional Support and Case Finding.'" *Translational Behavioral Medicine*, November, ibaa115. https://doi.org/10.1093/tbm/ibaa115.

FitzGerald, Chloë, and Samia Hurst. 2017. "Implicit Bias in Healthcare Professionals: A Systematic Review." *BMC Medical Ethics* 18 (1): 19. https://doi.org/10.1186/s12910-017-0179-8

Fox, Renée C. 1998. *Experiment Perilous: Physicians and Patients Facing the Unknown*. New Brunswick, N.J., U.S.A: Transaction Publishers.

Gallegos, Francisco. 2021. "Affective Injustice and Fundamental Affective Goods". *Journal of Social Philosophy* n/a (n/a). https://doi.org/10.1111/josp.12428

Halevi Hochwald, Inbal, Daniella Arieli, Zorian Radomyslsky, Yehuda Danon, and Rachel Nissanholtz-Gannot. 2022. "Emotion Work and Feeling Rules: Coping Strategies of Family Caregivers of People with End Stage Dementia in Israel—A Qualitative Study." *Dementia* 21 (4): 1154–1172. https://doi.org/10.1177/14713012211069732

Halpern, Jodi. 2001. *From Detached Concern to Empathy: Humanizing Medical Practice*. Oxford: Oxford Univ. Press.

Hamberg, Katarina. 2008. "Gender Bias in Medicine." *Women's Health* 4 (3): 237–243. https://doi.org/10.2217/17455057.4.3.237

Herron, Rachel V., Laura M. Funk, and Dale Spencer. 2019. "Responding the 'Wrong Way': The Emotion Work of Caring for a Family Member With Dementia". *The Gerontologist*, May, gnz047. https://doi.org/10.1093/geront/gnz047

Hoffmann, Diane E., and Anita J. Tarzian. 2001. "The Girl Who Cried Pain: A Bias against Women in the Treatment of Pain." *Journal of Law, Medicine & Ethics* 29 (1): 13–27. https://doi.org/10.1111/j.1748-720X.2001.tb00037.x

Holden, Kisha B., L. Dianne Bradford, Stephanie P. Hall, and Allyson S. Belton. 2014. "Prevalence and Correlates of Depressive Symptoms and Resiliency among African American Women in a Community-Based Primary Health Care Center." *Journal of Health Care for the Poor and Underserved* 24 (4A): 79–93. https://doi.org/10.1353/hpu.2014.0012

Lavallee, Zoey, and Anne-Marie Gagné-Julien. 2024. Affective injustice, sanism and psychiatry. *Synthese* 204: 94. https://doi.org/10.1007/s11229-024-04731-8

Lippi, Donatella, Raffaella Bianucci, and Simon Donell. 2020. "Gender Medicine: Its Historical Roots." *Postgraduate Medical Journal* 96 (1138): 480–486. https://doi.org/10.1136/postgradmedj-2019-137452

Manne, Kate. 2021. *Entitled: How Male Privilege Hurts Women*. London: Penguin Books.

Pismenny, Arina, Arina Pismenny, Gen Eickers, and Jesse Prinz. 2024. "Emotional Injustice". *Ergo an Open Access Journal of Philosophy* 11 (0). https://doi.org/10.3998/ergo.5711

Samulowitz, Anke, Ida Gremyr, Erik Eriksson, and Gunnel Hensing. 2018. "'Brave Men' and 'Emotional Women': A Theory-Guided Literature Review on Gender Bias in Health Care and Gendered Norms towards Patients with Chronic Pain." *Pain Research and Management* 2018: 1–14. https://doi.org/10.1155/2018/6358624

Schiavon, Cecilia C., Eduarda Marchetti, Léia G. Gurgel, Fernanda M. Busnello, and Caroline T. Reppold. 2017. "Optimism and Hope in Chronic Disease: A Systematic Review". *Frontiers in Psychology* 7 (January). https://doi.org/10.3389/fpsyg.2016.02022

Schulz, Richard, and Paula R. Sherwood. 2008. "Physical and Mental Health Effects of Family Caregiving." *AJN, American Journal of Nursing* 108 (9): 23–27. https://doi.org/10.1097/01.NAJ.0000336406.45248.4c

Sevelius, Jae M., Luis Gutierrez-Mock, Sophia Zamudio-Haas, Breonna McCree, Azize Ngo, Akira Jackson, Carla Clynes, et al. 2020. "Research with Marginalized Communities: Challenges to Continuity During the COVID-19 Pandemic." *AIDS and Behavior* 24 (7): 2009–2012. https://doi.org/10.1007/s10461-020-02920-3

Sullivan, Amy Burleson, and Deborah Miller. 2015. "Who Is Taking Care of the Caregiver?" *Journal of Patient Experience* 2 (1): 7–12. https://doi.org/10.1177/237437431500200103

Turner, J. 2000. "Emotional Dimensions of Chronic Disease." *Western Journal of Medicine* 172 (2): 124–128. https://doi.org/10.1136/ewjm.172.2.124

Wardrope, Alistair. 2015. "Medicalization and Epistemic Injustice." *Medicine, Health Care and Philosophy* 18 (3): 341–352. https://doi.org/10.1007/s11019-014-9608-3

Woods-Giscombe, Cheryl, Millicient Robinson, Dana Carthon, Stephanie Devane-Johnson, and Giselle Corbie-Smith. 2016. "Superwoman Schema, Stigma, Spirituality, and Culturally Sensitive Providers: Factors Influencing African American Women's Use of Mental Health Services." *J Best Pract Health Prof Divers*, 2016, 9(1) edition.

Part II
Affective Resources, Opportunities, Freedoms, and Recognition

In the book thus far, I have described in what ways patients, families, and healthcare providers may be subject to affective injustice in healthcare. I have shown that downgrading, ignoring, or silencing affects can have serious repercussions for healthcare actors and cause them harms that will affect their well-being as well as their capacity as knowers. I have also shown its negative effects on quality of care. My discussion of affective injustice has been closely related to the epistemic injustice discussion, in particular by showing how discounting affective knowledge can both harm the individual person, but also institutional learning and care outcomes. Given the seriousness of this problem, the question then becomes, what can we do about it?

The benefit of the epistemic injustice discussion from my perspective is that those who take it seriously, while understanding its theoretical richness, are not content for it to "just" be a theoretical concept and stay within the academy. This is largely because it is a non-ideal theory, rooted in attention to existing inequalities, domination, and structural oppression. According to Miranda Fricker (2017), and likeminded non-ideal theorists such as Ingrid Robeyns (2008) and Charles Mills (2005), it is only in clearly conceptualizing situations of injustice that we can understand what justice should look like; it is only in understanding what has failed that we will be ready to identify what we are seeking. Indeed, Fricker says her theory is both non-ideal and transitory (2017). In this way, she adopts a flexible approach, one in which scholars can engage in a continued exploration of inequality in a non-ideal world or work toward a more ideal theory by identifying remedial strategies.

In the themes that will be developed in the second part of this book, I will be engaged in this transitory approach by looking at those methodologies that help work toward an ideal of affective justice for healthcare actors. By affective justice, I mean a world in which affects are recognized, heard, and understood by healthcare actors and healthcare institutions. The goal of the second part of this book is therefore to describe promising methodologies that may be used, including a critical reflection on their possibilities and limits. Following Gallegos (2021), what I am interested in is providing affective resources, opportunities, freedoms, and recognition for healthcare actors. This includes:

–Affective resources and opportunities: activities, spaces, and circumstances that may contribute positively to one's subjective well-being (including providing self-care and safe spaces)
– Affective freedoms: freedom from interference in the pursuit of subjective well-being (such as avoiding emotional distress)
– Affective recognition: respectful consideration of, and responsiveness to, one's particular needs (and a recognition that one's affects are legitimate)

The following chapters will explore how to work toward these goods through methodologies that may be developed in hospital spaces and in medical training. Proposals will include a discussion of narrative methods, spiritual care, hospital design and architecture, inclusion of patient research partners, and pedagogies in medical education to give healthcare actors' affects a greater space. However, it must be stated that none of the approaches I will describe will be "enough" in isolation. Instead, they form a whole that helps work toward a change in the hospital, one in which healthcare providers, patients, and families can express and recognize their affects and those of others. As I will show, such affective goods will benefit persons as individuals, but also their relationships with their colleagues or patients. They will also benefit care provision by working toward more fruitful doctor-patient, colleague-to-colleague, and healthcare provider-institution relationships. They will also provide knowledge important to care decision-making, including better attention to the needs and values of patients, and in particular toward those who are currently marginalized in care provision.

This section therefore takes a transitory approach toward affective justice in healthcare, with a discussion of possible methods that will help healthcare actors' affects be better heard, understood, and recognized. They are meant to be the start of the discussion, the proposal of promising strategies that can be developed, tried, and tested to work toward the ideal of affective justice for all.

However, before continuing to these strategies, it will be important to recognize that in today's healthcare climate, there are multiple problems facing healthcare institutions. The first is healthcare provider burn-out, including the worrying increase of healthcare professionals leaving their professions post-COVID due to a loss of meaning in the profession and unacceptable working and pay conditions. The second is ever-increasing healthcare costs—both in light of an increasing aging population in many countries, but also skyrocketing costs due to technologies' increasing importance in care—meaning that healthcare is becoming more expensive, along with increased political and social pressure to reduce healthcare costs. This means that healthcare systems in many countries are at their breaking point. Therefore, some of the strategies I will talk about in Part II may appear ideal: giving space to narratives, taking time for spiritual care, investing in hospital design and architecture, imagining ways to integrate patient experts as an affective resource for patients; however, my strategy in Part II is based both on an (ideal) and a reality of what may be feasible if we can accept affects as an important part of healthcare work. First of all, to get an idea of what affective justice may look like, we will need to at least be able to imagine this ideal, and for this to happen, we need to have an understanding of what resources and opportunities we need to reach it. However, given increased resource shortages, we also need to be realistic. Therefore, most of the strategies I propose are already in place and can be adapted, or if they are not (yet) in place, they can be put into place with minimal costs. As I discussed in the last chapter (Chapter 3), a phenomenological toolkit can be used in staff meetings or in trainings with little cost or time taken from care. Similarly, recognizing the spiritual care provider as an affective resource means valuing their role, but not necessarily more time or resources to achieve it if the provider is already integrated into the service. While some of these affective resources may require more time/hospital investment (such as narrative medicine workshops), I will discuss strategies (such as the PENbank) which have been implemented in hospitals with little cost or time commitment thanks to small incentives given to healthcare providers. Finally, I have been arguing throughout Part I that taking seriously affects and their importance to quality care is also a wise investment for hospitals in terms of staff retention and patient satisfaction. For this reason, they can be considered not only ideal but also having a practical aim of improving care quality. However, in order for these strategies to be convincing to hospital management, as researchers we can also work toward this goal by helping to gather patient and healthcare provider perspectives of these resources and opportunities and investigating their link to healthcare provider and patient well-being and quality care. It is in this way that we can turn these ideals into real strategies and investments by the hospital and move toward affective justice for healthcare providers, patients, and caregivers.

References

Fricker, Miranda. 2017. "Epistemic Contribution as a Central Human Capability". In *Equal Society—Essays on Equality in Theory and Practice.*

Gallegos, Francisco. 2021. "Affective Injustice and Fundamental Affective Goods". *Journal of Social Philosophy* n/a (n/a). https://doi.org/10.1111/josp.12428

Mills, Charles W. 2005. "'Ideal Theory' as Ideology." *Hypatia* 20 (3): 165–183. https://doi.org/10.1111/j.1527-2001.2005.tb00493.x

Robeyns, Ingrid. 2008. "Ideal Theory in Theory and Practice:" *Social Theory and Practice* 34 (3): 341–362. https://doi.org/10.5840/soctheorpract200834321

Chapter 4
Affective Justice Through Narrative Methods

Abstract This chapter will look at the possibilities and limits of narrative methods to help work toward affective injustice for healthcare providers. In a first introductory section, I will review the debate on narrative methods to understand the context in which they have been promoted and their limitations. I will then investigate the ways and means narrative methods may provide affective resources and opportunities for healthcare providers. In this chapter, there is a dual goal I am seeking in promoting narrative methods. This includes improving the well-being of workers, but also improving care, in particular by promoting alternative means of institutional learning and reporting. As this is a large debate, I have chosen to focus on four narrative methods that are particularly promising for these two ambitions: (1) free-form journals written by healthcare professionals; (2) narrative medicine seminars in the hospital; (3) recording and safekeeping healthcare providers' affects through narrative banks; (4) collective narratives gathered during interdisciplinary reflection and analysis of practice. Each of these methods will be analyzed and their possibilities and limits explored in the context of affective justice.

4.1 Introduction

The stories of patients and their lives, in some of their worst and best times, lived with and through their emotions, has long inspired healthcare providers to reflect on their experiences of clinical work. From the vivid case studies of Olivier Sacks (1998) to fiction-writing (Verghese, Doyle, Chekhov, Hossini, etc.), healthcare providers rely on writing and telling to better understand, apprehend, and make sense of their experiences, as well as to share them with others. As the French philosopher Paul Ricoeur has said, narration can help a person find one's place in the world (2006). By telling, by being told, by being able to say, we start to recognize ourselves as capable and become actors in the stories of our lives. Narrative methods may empower the individual—and in our case, healthcare providers—and help them to better face the challenges of medicine. They also may feed into institutional knowledge and change.

B. Bogaert, *Affective Injustice in Healthcare*, The International Library of Bioethics 112, https://doi.org/10.1007/978-3-031-94378-2_4

In line with Gallegos' proposal (2021, p. 189), what I am seeking in this chapter is promising tools that permit a certain number of affective goods (freedoms, resources, opportunities, forms of recognition) for healthcare providers. In this chapter, I will show in what ways narrative methods may facilitate access to certain affective resources and opportunities and types of affective recognition. By affective resources and opportunities, I will look in particular at those activities, spaces, and circumstances that contribute to the healthcare provider's well-being. By affective recognition, I will investigate to what extent narrative methods may help healthcare providers be better recognized as affective beings by other healthcare workers and their institutions.

My purpose in this chapter is double: (1) to look at how narrative methods may help the healthcare provider to experience greater well-being by providing a safe space to express their affects; (2) to understand how these methods permit affects to become part of the knowledge base of healthcare by providing new information on healthcare work (a counter narrative) that is inaccessible via other reporting measures. In this chapter, we will see that promoting alternative methods in healthcare institutions such as narrative banks, workshops, and writing may enable institutions to become more responsive both to the affective needs of healthcare providers, but will also enable the institution to gain information not available through other, more formalized methods. In other words, I am interested in both the goals of affective and epistemic justice being achieved through narration, particularly when we can recognize these affects as forms of knowledge important for care provision. This second point is an important one for our subject. In the private relationship between a reader and a writer that takes place through the writing and reading process, we might be tempted to regulate narratives to a private sphere and to focus exclusively on how reading or writing narratives might improve the healthcare provider's well-being. While this is a promising means of ensuring that healthcare providers have an outlet to express themselves without constraint or fear of hierarchies, limiting their potential to the private domain will do little to value affects in healthcare institutions. As I defend the idea of affects as forms of knowledge, they will need to be taken seriously at a higher level and guide institutional knowledge-making. Only then will the hospital become a safe affective space for all.

In this discussion, I have chosen to discuss several promising methods to show the possibilities and limits of narrative methods to work toward affective justice for healthcare providers. The first is a proposition of free-form journals written by healthcare professionals to provide them with an affective resource turned toward their relationships with patients, but also their own affective needs. The second is more well-known and practiced: narrative medicine groups. While these types of initiatives are being implemented by researchers and literacy scholars around the world, I have chosen a case study to understand what happens when an institution explicitly supports and advocates for it in the hospital space. The third, still in its phase of experimentation, seeks to collect anonymous healthcare provider narratives and safeguard them in a bank, as well as use for them for institutional learning. The fourth case study takes a different track and looks at how collective narratives may be valued in interdisciplinary meetings. This case study is quite different in character—as it involves oral narrations that may or may not be written, but as I will show, they can also create counter-narratives of care. Once I have outlined these various initiatives, I will conclude by looking at the various possibilities and limits of working toward affective justice through narrative methods.

4.2 Free-Form Journals: A Safe Space for Affective Expression

The first resource is the possibilities offered by free form journals. By free-form journals, I am referring to the possibility for healthcare workers to document their care experiences and other work experiences by writing down their experiences in a free format in a journal. They are an open method, in which the writers may choose a format that is available/feasible for them, including writing their experiences as stories (whether factual or fictionalized), as summaries of their daily/weekly/regular experiences, as bullet points, or as spaces to express key emotionally charged moments (such as the death of a patient or a conflict with their colleagues). In practice, many healthcare professionals keep personal journals such as these, although they are mainly meant for personal use and are not directly supported by their institutions or written during their worktime.

The source of my analysis comes from a research project that analyzed free forms journals written by health and social care professionals during the first wave of the COVID crisis in France.[1] The overall goal of the project was to use narrative methods to better understand what happened during the crisis in these care structures and to benefit from their experiences to improve crisis planning and care provision. Professionals from social-medical structures in eight French regions were invited to reflect on their experiences of care provision during the crisis through these journals. They were solicited to participate from a coordination body that supports these structures in France (CREAI). Professionals were allowed to write their journals during regular office time; however, due to the realities and complicated organization during the crisis, they also often wrote them in their free time. The overall project had the support of management in each of the participating structures; however, professionals were told that their journals would not be read by their hierarchies and only used by researchers. Thus, we can assume that care providers knew their journals would be read, but not by their managers, given them a certain amount of free expression about the realities of care provision during the crisis (but with a reader in mind) (Table 4.1).

[1] This research project was supported by the French funding agency (ANR) and lead by an association that overseas medical and social structures in France (CREAI). While other publications have resulted from this research project (see in particular Bogaert and Pierron 2022, which discussed the healthcare values put into focus in these journals), the analysis proposed here is new. It is also to be noted that the excerpts presented in Table 4.1 were translated from French–English.

Table 4.1 Excerpts from journals classified by affects

Emotion	Excerpts from journal
Fear	F1: The virus was everywhere, and that's where, in our fears, the personal and the professional came together, sometimes so brutally that it could be very complicated to sort things out and be reassured enough to be able to come to work calmly and go home at peace F2: I'm afraid she'll feel lonely or abandoned (I call her every week) (the writer is referring to a colleague)
Anxiety	A1: We're not heroes…we fight our anxieties in different, singular ways. We realize how anxiety can paralyze some people, psychically paralyzing them. We need to remain firm on the framework, on principles, on the common interest, but we need to reach out to those who feel lost, welcoming them where they are distressed A2: This time, I had to deal with the anxiety of a colleague who gave me an idea of the insecurity that this crisis could generate in others….this realization was painful, but at the same time led me to constantly question my posture and the way in which I should henceforth organize my work in line with collectively shared protocols
Anger	AN1: It's been a hard day. 9:00 pm, I send a message to my two colleagues. I tell them how tired I am, both physically and mentally, and how sorry I am at the way my grief has taken on all too often: anger. The difficulty of this recovery and my despair at the test of individualism. I tell them that they count, that their presence is dear to me, not without a few tears AN2: Feelings of frustration, anger and injustice. Fatigue. Emotional outburst (I let go in the face of this wave and immersion of stress that I had tried to anticipate), then step back: we're managing, but once again in a hurry when we could have foreseen it, but we're going to manage it
Sadness	S1: But let's go back to last night…to our management meeting. We were able to express our sadness and anger at seeing so little generosity on the part of some. Only some. At first we said "the team." But no, it's just a few who spoil the mood. The crisis brings out the best and worst in everyone. And that's no surprise S2: Since her arrival last October, we've had very little to say about educational support…discussions have been reduced to managing the organization of the department. I feel that my position has been considerably impoverished, and that saddens me deeply S3: I know this young man is already struggling at school. The situation is only going to get worse, which makes me sad or bitter. In the weeks to come, I'm not going to be able to stand by and do nothing…
Happiness	H1: I'm happy to see everyone again. I'm warmly welcomed by everyone…they ask me if I've had a good rest, if it's not too hard to "get back to work" H2: But also, I'm happy to see that everyone is finding their feet again and trying to take care of things in their own way. A collective that's starting to take shape again

The professionals who agreed to participate came from a diversity of social and health structures, including those working in child and elderly protection to specialized services such as addictology. The writers included directors, psychologists, nurses and care assistants, secretaires and maintenance staff, thus representing a diversity of care professionals. More women (47) compared to men (18) participated, which is a reflection of the majority of female staff working in these structures.

The journals varied considerably in terms of writing style, format and theme, from simple entries relating daily activities to retrospective pieces written after the high point of the crisis. They often included hour-by-hour or day-by-day commentary relating the changes put into place (most often a transition from in-person to remote care), how their patients/users were affected by them, and what they thought and believed about the choices being made. Others—due to time constraints during the crisis or simply because they preferred to reflect a-posteri, used the journals as a means to consider the changes they had put into place, their capabilities to act, and its effects on their users. This narrative method therefore provided a reflective space, one that encouraged care workers to think about their role in the crisis and how it affected them and their users.

To better understand how these journals functioned as affective safe spaces, the table below shows a few excerpts, chosen based upon the affects most commonly expressed in the journals: fear (91 times), anxiety (61 times), anger (27 times) happiness (19 times), sadness (18 times).

From this table, we can see a snapshot of the daily lives of these careworkers during the pandemic, in their stressors and anxieties, in their relief and sadness, in their anger and fear, related to the reconfiguration (and sometimes the breakdown) of the care plan for vulnerable users during the crisis, the engagement (or disengagement) of their colleagues, and the comprehension (or miscomprehension) of their hierarchies and governments to support vulnerable persons.

As we can see from these entries, professionals used these free-form journals as a reflective space to think about the crisis, their roles as care professionals, and their relationships with their users and colleagues. Their ways of expressing affects included both witnessing them in others (their users, their colleagues) and telling of their own. While some entries were related to the specificities of the sanitary crisis (such as fear of contamination), the objects of these emotions were mainly focused on their relationships with their users and with fellow workers. The disconnect between an ethics of care and the care organization put into place during the crisis was the most frequent theme/entry in the journals. It led to feelings of anger, frustration, fear, and sadness.

As we can see from the excerpts, the free-form journals also gave a free space for the expression of strong emotions, such as fear or anger, which were narrated without constraint (and often specifically directed toward colleagues or their relationships with their hierarchies). Writing these negative affects down was a means of letting go, of releasing tension; however, they also served a reflective purpose. As we can see from the entries, they also permitted a reflection on the object of the emotion, which was often related to how they felt about their work (their relationships with their users/beneficiaries and with their colleagues). The themes described included:

1. Feelings of anger and injustice relating to a care organization that is not personalized to the needs of vulnerable users
2. Anger and sadness about a management who "hears" them but does not "recognize" them

3. Feelings of desperation, hopeless, fear, and sadness about the lack of team communication and coordination
4. Happiness to be able to do their jobs, in particular in the field and in contact with vulnerable populations, as well as to work together in a team when it was possible

As we can see from these descriptions, the free space provided by this narrative method put into focus their care values. Among reasons for feeling and expressing these affects, we can see the lack of a care plan adapted to vulnerable users and the difficulties with hierarchies and team organization. The narrative method allowed these issues to come to a focus for themselves and to witness and question their roles and their possibilities for individual action in a constrained environment. The format of the journal provided a safe space for expression, including of their affects, their values, their beliefs about their profession, and the (too frequent) disconnect between what they valued and what they practiced. This affective resource both helped professionals to tell but also to become more aware of their values.

As we can see, this method fulfills the first goal of affective justice by providing a safe space/resource for healthcare professionals, which in turns helps foster their well-being. It provides a space to "put down" their emotions, to release tensions, to reflect on what it said about them and their professions. Although more research is needed to understand if this method had an impact post-crisis and was continued by care providers of their own initiative or with the support of their institutions, from the entries, we can already see its value as an affective resource for care professionals during the crisis.

In addition, these journals also serve as a "counter narrative" that is quite different in form and content from other reporting measures these providers habitually use, in particular the patient/user file. This free form of reporting is also turned toward the vulnerable user and their needs, but it is also focused on how professionals live care work, including their relationships with colleagues. It therefore provides access to knowledge that is not written down elsewhere. We can thus consider these free-form journals as a type of knowledge gathering and reporting that might be valuable to share at a higher level. In addition, the objective of the research project was to be able to learn from the crisis, directly from care professionals. This suggests that they will need to be shared with hierarchies and institutions in some way, if we take seriously the proposal that affects (and other themes in the journals) say something important about care work.

As discussed at the beginning of this case study, the research project guaranteed autonomous participation of care workers. The journals themselves were only shared and read among the research team and all narrations were anonymized in the analysis of the research results to prevent identification. This was the condition that enabled the safe space for participants to voice their affects openly, notably in discussions about relationships with colleagues or with management. While the articles/report produced may guide institutional learning, in particular as the support body (CREAI) reported back to the teams involved, there was no direct connection between the writer and the hierarchy. The "reteller" of the story was the "reader/researcher" whose job it was to voice these experiences and have them be heard. This means that care providers did not really have the capacity to act on their own stories, nor to decide which stories were the most meaningful. This means that their narratives largely remained in their private domain. This is not necessarily a problem, if our focus is on providing an individual affective resource; however, from a practical (and organizational) angle, it means specific needs and problems could not be dealt with by individual structures. It also did not help move toward a change in the hospital, one in which affects could be better valued in teams. Because of this, while this type of method may provide an individual affective resource for healthcare workers, it has a limited impact to improve and reflect on care.

The ambiguous place for these forms of reporting also raises an important question on the value and place for counter narratives such as these in care institutions. As Paul Ricoeur (2006, p. 152) poignantly says, *"learning to tell your story is also learning to tell it differently."* By providing a free space to express their affects—including their critical feelings about their care structures and how they were run—these journals created a counter narrative in opposition to other reporting measures, which largely consist of case files written about the patient. These case files rarely reveal institutional problems, at least explicitly. The journals on the other hand gave a different story about what happened during the crisis, one that could even be said to be dramatically opposed to what was reported in the case file. Because of this, free-form journals such as these may create conflict. In this research project, one care structure continued their journals beyond the crisis and decided to share them as a team: the result was to create tension, as the journals were a radically different means of reporting on care work and determining what was important. This first discussion therefore shows the caution that needs to be taken when we try to move from an individual affective resource to a collective one, particularly when the institution is not ready for such alternative perspectives.

However, despite these limitations, free-form journals do present a promising resource that allows affects to be given a greater place. The kind of reflection permitted by the journals is rare, as there are few spaces in care work for such kinds of reporting. As our discussion has shown, they largely have the potential to be used as individual affective resources to help the care provider to reflect and express themselves without constraint. As they may help in this way to foster well-being, we have already gone quite a way with our first ambition in the chapter by providing an affective resource to be used by care workers. However, the place of narrative methods relative to our second ambition, which involves improving care and promoting institutional learning, is far from resolved. Let us therefore turn to a second example to see what happens when an individual narrative is moved to a safe space in a collective.

4.3 Narrative Medicine Workshops

This second section will discuss the potential and limitations of using narrative medicine workshops as a safe space for affective expression of healthcare workers. I will first briefly introduce the narrative medicine movement and its ambitions before moving on to a specific case study that appears particularly promising to our debate. The reason I have chosen to discuss this initiative in particular is that their narrative medicine workshops have been financed and encouraged by the hospital, meaning that narrative methods have been given explicit support by hospital management. It will therefore allow us to explore whether or not they can fulfill our two ambitions turned toward affective justice.

To begin with, narrative-based methods in clinical practice has long been practiced by doctors as a way to cultivate a better relationship with their patient and to improve care. However, since the second world war, such methods were (partly) regulated to the background due to the promotion of evidence-based methods. As we have already discussed, these methods favored science-based practices based upon "facts and findings." In this light, these new methods were seen as objective and scientific compared to the more subjective experiences developed through patient narratives (Kalitzkus and Matthiessen 2009). The value of narrative methods to improve clinical practice has however regained interest in the past decades. Interest in narrative ethics (Ricoeur and McLaughlin 2009) and the emergence of the narrative medicine movement (Charon 2008) have helped bring back attention to its value in clinical care. The narrative medicine movement as it is known today started in the late 1980s and gained increased popularity in the early 2000s, with seminal works by Arthur Kleinman (1989) and Rita Charon (2008). These works have brought anthropological and philosophical perspectives to the understanding of illness experience and helped put to the forefront the importance of patient stories to improve medicine and practice. Charon (2008, p.4) defines narrative medicine as, "narrative skills of recognizing, absorbing, interpreting, and being moved by the stories of illness." As can be seen from this quotation, the wording of narrative medicine was coined to counteract the shortcomings of evidence-based medicine (Kalitzkus and Matthiessen 2009). While this could be seen as a critique of EBM, the word that Charon uses in her definition is "strengthen." As a doctor, she is not denying the importance of EBM-based practices. What she is trying to achieve is to improve them through the reinforcement of narrative methods.

By refocusing attention on the narrative of the patient, narrative medicine's goal (at least in the formulation of Charon) is to help the healthcare provider gain a more global vision of the patient, including being "moved" by their stories. The assumption behind the movement is subjective views (including affects) revealed by narrative methods will allow the clinician to explore dimensions inaccessible with other methods. While there is no consensus on a definition of what exactly narrative medicine is, most models enable the clinician to gain more information about the patient (including their values, beliefs, and emotions). They aim toward person-centered care, better patient-clinician relationships, and improving diagnosis and treatment. They have also inspired complementary and parallel care movements, including Gothenburg's person-centered care model (Ekman et al. 2021), which has made narration part of a three-part process to work with the patient.

How can narrative medicine achieve these ambitions? Due to current gaps in medical training in developing narrative competency, the movement has advocated that we need to train doctors to give greater attention to patient narratives in clinical practice. This has led to a diversity of initiatives, but most often they take the form of narrative medicine workshops/trainings. Most initiatives are based upon or inspired by Columbia University's models of training courses incorporating three main activities: (1) attentive reading; (2) expressive, reflective, or creative writing; (3) sharing (Delorenzo et al. 2021). A workshop normally consists of reading and analysis of a narrative, followed by personal writing inspired by the narrative and sharing these narratives with others in the group. Through these workshops, healthcare professionals will develop narrative competencies, enabling them to provide a greater place and space for patient stories in healthcare practice.

Experiences thus far suggest that narrative medicine workshops can be beneficial to care provision, notably by improving care skills, including relational and cultural abilities, and enhanced empathy (Remein et al. 2020) and that they can contribute to professionals' well-being and reduce burn-out (Bajaj et al. 2023). While Charon is careful to distinguish her proposal from those of other workshops focused specifically on affects (such as well-being workshops), they do allow the participants to explore and recognize their patient's emotions. This happens by providing clues and information not available through other methods. In particular, affects in patient stories may be seen as "clues," which may say something important about the diagnosis or treatment of the individual person. From this standpoint, they may be recognized as a form of knowledge for the clinician, or at least as a way to help understand the patient's story and what is of importance to them. While this may work toward affective justice for patients, in this chapter I will investigate what happens in these workshops for healthcare providers.

To discuss whether or not narrative medicine workshops may help move toward affective justice for healthcare providers, I will use an example of the narrative medicine workshops proposed since 2018 at a hospital in Paris. It is an initiative developed by a literary scholar in close mentorship with Rita Charon (Delorenzo et al. 2021). As already discussed, I have chosen this model as it has been explicitly financed by the hospital—including in terms of healthcare provider time (it is considered worktime). This case study will allow exploration of the second goal of this chapter, namely if such kinds of affective resources may be shared to promote institutional learning. It seems fitting to explore the possibilities and limits of the method in an institution that explicitly supports it.

Practically, the hospital organizes interprofessional groups that meet once a month to take part in a two-hour workshop. To encourage freedom of expression, the groups are composed of professionals who do not work in the same departments. In addition, participants are guaranteed confidentiality, in that the texts written during the workshops are only circulated within the groups. The method consists of three phases: (1) basic workshops; (2) creativity workshops; (3) co-facilitation workshops. This format closely follows the narrative medicine pedagogies proposed by Charon that consist of the steps of close reading, writing, and sharing, but they add an additional goal of trying to create a "narrative" culture at the hospital by training new facilitators in step 3.

The question important to our debate is the extent to which these methods give a safe affective space for healthcare workers to express themselves. First of all, as the purpose of the workshops is to develop narrative competencies and to provide another means of reporting/writing/telling different from a typical patient file, it holds similar possibilities with our previous case study. It provides a counter-narrative that may allow more free expression to emerge. According to the authors, "it is 'parallel' to hospital records. In a parallel file…in which caregivers have the possibility and the right to write, using ordinary language and/or the expressive form that seems most appropriate to them, what cannot be written in a classic file, but is necessary for care."[2]

Delorenzo et al. (2021) describe that in the first phase (close reading), the group members read closely a text and build up a representation of a narrative using imagination. This first phase reinforces affiliation amongst the group members, as they share aloud their responses to the text, and in particular to "our aesthetic and emotional responses" to it (Delorenzo et al. 2021, p. 49). In this phase, we can understand the ways in which affects come to the attention of the participants. First of all, in a narrative, there is a deliberate decision of the writer to express feelings, whether as a narrator or through a character in a story. These affects may be explicit, such as when the writer expresses what the character is feeling; however, it may also be more implicit, such as when a character's emotion is revealed through a narrated monologue or external cues such how the character dresses or acts (Keen 2015, p. 152). These affects are directed towards a reader in order to create a mood, a nuance, or an understanding of the character or the situation. In return, for the reader, affects function to help them empathize with the characters and to live through the story that the writer wishes to tell.

However, what happens when the reading moves from the individual relationship between the writer and the reader and into a collective space, as the narrative workshop proposes? It appears to provide a common, reflective space in which the person's affects can be shared with others and taken seriously. As Delorenzo et al. describe it, this collective sharing of emotions (and other responses to the text) reinforce affiliation amongst the group members and help them come together as a group. Here affects will take a central place in the telling of the stories, as group members describe their "emotional responses" to the text. In this way, they provide affective recognition (by the reader, who will take these affects seriously, as something important to be shared) and in the discussions of the text with other participants.

The authors go on to say that all of the steps in the process are interlinked and that "we bring new, more personal and expressive, creativity and creative attention to the original narrative" during the other phases of the process. In the next phase of the workshop (creative expression), the method involves having participants write

[2] Our translation to English. The article was originally written in French.

and share narrative pieces, which can range from real-life case studies to fictional narratives. We can expect here too for affects to have a central place in the telling and discussions, as they have learned from their close reading that they take an important place in writing process. In the final step, which involves having care workers co-facilitate workshops for new members, we can also expect that if the participants have learned from the safe affective space permitted by these workshops, the new trainer will also encourage this kind of expression in their workshops.

From this description, we can see affects play a role in each phase of the process: through the intermediary of reading, participants learn to express what they feel, affective capabilities that are further developed and encouraged in the next stage, when they become the writers and facilitators. A participant said that the workshops, *"made me think about my day-to-day posture as a carer…each life story helps us to understand how patients and their families feel at the time we take care of them, and how this affects us, or not…writing, during or outside the workshops, also became a kind of outlet, allowing me to express what was difficult for me to say orally."* As we can see from this testimony, the workshops functioned as an affective resource for healthcare providers to openly express their feelings on healthcare experiences, including finding ways to express things difficult to express in other ways. Affects became a part of reporting on care.

While this case study is similar to the affective possibilities provided by free-form journals, it does give at least a partial means of moving beyond the individual affective safe space to enable affective sharing among a group (albeit in an enclosed, restricted group). Compared to our example of free-form journals used for research purposes, it therefore holds more potential to guide change in the hospital about how affects are seen and understood. The possibility of expression without restraint is pivotal here to provide the safe space necessary. It was made possible by: (1) freedom from sharing to hierarchies and management; (2) support from the healthcare institution, in particular by making the option to participate part of their worktime.

Although in this example we have only partly moved from the individual affective resource to a (more) collective one, we can see its potential—if it continues to be supported by the healthcare institution—to gradually help affects to be more valued, as they are openly discussed and valued in these spaces. In the long-run, this may make it easier for narrations (and in turn affects) to be valued, at least among those trained via these workshops. While we have still not solved the problem of how these narrations may directly feed into reporting and institutional learning, they do show the possibility for them to enable a safe collective, reflective affective space in a small group.

The limitation of this method however is that affects are not necessarily made explicit and are only a secondary goal. This may mean that those who do not believe affects are important to care practice, or not necessarily aware of what role affects may have, still may not see them as an affective resource. Specific workshops on affects through close reading and writing may help to bring attention to this question and enable them have a more direct impact on care practices. Now having explored the potential and limitations of narrative medicine workshops, let us now turn to a third initiative to better understand how the use of narrative methods may help work toward our two affective goals.

4.4 Narrative Banks: Anonymous Affective Safekeeping and Sharing

A third method to be explored is narrative banks. Here I will take the example of the PENbank, an initiative of the Lausanne University Hospital to gather and safekeep anonymous, written experiences of healthcare providers. Inspired by the idea of biobanks, its goal is to be able to collect, store, and use doctor's experiences to improve care. As I will discuss, the advantage of this resource is that it allows the healthcare provider's affects to be shared with a wide audience, including at an institutional level, thereby working toward the second goal in affective justice.

To start with, the PENbank is an infrastructure at the hospital created in 2021 that collects, preserves, and displays the lived experiences of healthcare workers (CHUV 2024a). It is a "bank" that safekeeps stories of doctors and their practices of medicine. The idea is both to keep these narratives safe in a secured database, to use them for research purposes, but also to have information contained therein to reach management. Like our example of free-form journals, these testimonies can take a variety of forms, from oral testimonies to written ones. They include not only stories/narratives, but also photographs and drawings. The initiative motivates healthcare providers to participate by giving small gifts, such as free coffee, as well as holds contests for the best entries. The narratives are publicly displayed at the hospital and regularly shared in the hospital's monthly newsletter, which allows all hospital staff (doctors, administrators, researchers) to be able to read the best entries. The following is an example of narrative of a pediatrician from the PENbank, which was shared in the hospital-wide newsletter in 2024:

> Relief, Joy.
>
> Oh, how far away that delicious day seems already. You, wriggling in front of me from the height of your 4 hour life, who cares if I ticked the right boxes that day, sitting on my bottom in a crowded auditorium of feverish aspiring doctors?
>
> You were able to breathe on your own as soon as we took you out of the warm womb, and for that I thank you. I was far more relieved than you were, I promise you, when I heard your crisp, clear cry. And now you'd just like me to return you to your mother's breast, instead of trying to measure your head circumference or check that you're spreading your arms wide when I tip you backwards. Don't worry, I'll have you back in no time.
>
> I have a badge that says I'm a Doctor, and a phone in my pocket that rings when you need a pediatrician in the delivery room.
>
> Never has my confidence in HR's competence been so shaken as the day they entrusted me with this responsibility (and yet, oh how fragile it was).
>
> I inspected you from head to toe to make sure nothing was missing, nothing was wrong.
>
> You can't imagine how good these few minutes with you, in the quiet of the examination room, make me feel. I place you gently back in your little bed, and fill myself with all the

miracle you embody with your 3200 g of life. To return to the whirlwind of those first days of labor.

Story sent by a doctor, May 2023.[3]

As we can see from this entry, the written (and anonymous) form gives the person the opportunity to freely express their affects (in this case their joy, relief) about a healthy baby delivery. However, we can also see their anxiety about doing well for their patient *("Never has my confidence in HR's competence been so shaken as the day they entrusted me with this responsibility (and yet, oh how fragile it was")*. Thus, while the entry is on the surface about a happy event, it also shows the anxiety a young doctor feels in having the responsibility of a new life in their hands.

As we can see, the narrative bank functions in a similar way to free-form journals or narrative medicine workshops by providing a space in which affects are visualized and understood as part of care experiences. However, compared to our previous example, the narrative bank is made available to a wider group. First of all, there is the possibility for them to be read by any hospital personnel, either via the newsletter or by printing of the stories that are displayed at the hospital. The second way they are read is by researchers, who use them for their own analysis and research priorities.

The third ambition however is to use them as institutional resources. Researchers do this by giving annual feedback to the institution's governing bodies via a "map" that thematizes what these stories express. The organizers state that, "knowing how they face the challenges of the clinic opens up ways of supporting them and raising awareness of these issues among hierarchies…it is a question of paying attention to an essential player in the healthcare system (Le projet PenBank 2024)." Knowing what doctors experience enables them to better understand what is important to them and what preoccupies them (Bourquin et al. 2022). We can therefore understand that if affects are openly expressed in these narratives, it may be a means of enabling the healthcare institution to profit from this affective knowledge, or in the very least to pay more attention to it in their work. However, the risk of taking these narratives at face value is that the writers "write with a reader in mind." By knowing that it may be shared at a hospital-level, it may encourage some persons to focus more on positive emotions (joy, relief). However, as we can see from the example shared above, while the doctor largely focused on happy experiences, it also—in a more subtle way—provided a space for the healthcare provider's anxieties to appear.

As we can see, the advantage of this method, in comparison to the previous two methods we have discussed, is these narratives can be read and taken seriously outside of an individual or small group. While the initiative does not show the place of affects in the annual feedback report, we can expect it may become a central theme in the telling of healthcare. The initiative was first based upon a pilot test during the Covid crisis from April-June 2020 in which around 30 written or audio-recorded accounts, drawings, and photographs were collected. The researchers developed a

[3] Our translation from French—English. This story was published in the internal newsletter of Lausanne University Hospital (Infine), making it widely available to healthcare staff. Stories from the PenBank are published every month in this newsletter (for this particular story, see: CHUV Infine January 2024b).

number of themes based upon these entries. An important theme to emerge was the ways participants encountered and experienced the crisis in emotional terms (Bourquin et al. 2022). Thus, we can expect affects to take a place in the thematization done by the researchers, when they are prioritized by its writers. In addition, the "rough/full" narrative, as it is made visible in a variety of formats, can also be read by hospital workers/managers, enabling them also to have access to the data beyond the researchers' interpretations. Because of this, they represent a promising possibility as affective resources and opportunities at the institutional level, including by contributing to a change in perspective among healthcare workers, one in which affects may come to have a real place.

On the other hand, as these narratives are anonymous, they do not benefit from collective exchange and reflection. This resource is therefore limiting in the same way as the passive affective sharing in our free-form journals. It also does not seem to be the means to change the individual care worker's reluctance to share affects in their teams. Like the free-form journals, it also has the disadvantage of being created by a writer who knew they would be read/seen, and therefore could use affects for other purposes than the retelling of healthcare. Because of these limitations, let us turn to a final narrative method, one focused on gathering collective narratives in a team.

4.5 Narrating Cases: Interdisciplinary Reflection as a Collective Affective Resource

In this final section, I will explore a very different means of narration, namely narration by healthcare providers about patients when discussing case files. As this type of situation takes place "aloud" and in a collective manner, it will be very different from our previous examples, although the narrative medicine workshops did provide some possibility for sharing orally, albeit in a small group. In addition, this resource has a fundamentally different focus. While the narrative methods discussed thus far are largely introspective, centered on narrating healthcare professional experiences, these types of groups are geared directly toward and for the patient. However, I advocate that this resource may be promising by allowing a safe space for affective sharing among colleagues, as well as leading to some institutional learning.

There are two principal methods used for collective reflection among colleagues in hospital spaces today. The first is interdisciplinary reflection on difficult cases. The purpose of these meetings to reflect and discuss collectively about difficult patients in their care and to come to a common decision. Difficulties in the care plan come from conflictual patient and family relationships; when "no good" care solution may be found due to financial or administrative barriers; or simply due to the limits of medical knowledge and treatment. These meetings are important both for the doctor (responsible for the case) who can benefit from collective reflection (and not make the decision alone), as well as giving the possibility for care workers other than the doctor (the psychologist, the nurse) to have their knowledge be heard and understood. This is important in particular for professionals who are closer to the patient (such as nurses) but are often left out of collective decision-making, which makes them more prone to both epistemic injustice and moral distress (Reed and Rishel 2015).

A second method is analysis of best practice, which are sessions organized in particular for young doctors to benefit from collective reflection among peers. At times, they may involve other care professionals, but they are often limited to young doctors—meaning other care workers, such as nurses or doctors with more professional experience—are not invited. These spaces are meant to provide a safe space for these young care workers to be able to express themselves in a non-judgmental environment and to reflect "aloud" and to other young doctors what went wrong in a given case and find solutions for how to do better in the future.

These collective reflection spaces are complementary, but also radically different in their approaches. The first is geared toward making a decision on the patient (present tense) while analysis of best practice is retrospective (past tense). The first is geared toward benefitting from interdisciplinary expertise, while the second is geared specifically toward the challenges faced by one particular type of healthcare provider (young doctors). Finally, while the first involves professionals at various stages of their careers, the second is geared toward developing professionals' competencies among a younger generation.

However, they both can serve as affective resources. In the first case, it is believed that by benefitting from the experiences and expertise of each healthcare provider, the team is more likely to come to a collective decision that respects the values and beliefs of the patient. As each healthcare professional has a different relationship to the patient (in some cases this is very different, from cooperative to conflictual relationships), having each person share their visions, their experiences, and their expertise helps the healthcare team to better see and understand their patient in their globality and make decisions on their care. In the second case, analysis of practice often involves reflecting on cases in which something went wrong, such as a medical error, a bad outcome, or a conflict in the team. Analysis of practice not only helps take care of these young doctors by giving them a safe space to discuss what happened, but it also allows them to improve patient care by learning by their mistakes. I will argue that both methods can also provide a forum for healthcare provider's needs, and in particular their affective needs.

For the first example, I will use testimony collected from a research project conducted on interdisciplinary reflection for difficult cases in palliative care (*reunions pluridisplinaire de coordination*) in France.[4] The findings to be discussed here relate both to the ethnography stage, when researchers observed what happened during these meetings, and the second phase, in which we conducted interviews with

---

[4] The methods used in this study consisted of ethnography of online meetings for a period of three months, as well as retrospective interviews conducted with participating healthcare providers. The focus of this research was on how these meetings facilitated epistemic injustice for various categories of professionals in spite of hierarchies.

healthcare providers about the purpose of these meetings and their value to care practices.

In the interviews, healthcare providers stated the goal of these meetings was to help their patient, to find a path to the "least bad" decision by benefitting from each professional's expertise. The secondary goal was to work toward group solidarity and come together as a team. While not the explicit goal of these meetings, healthcare professionals interviewed believed healthcare providers' affects were an inevitable part of these meetings. The meetings were geared toward end-of-life, including the possibilities for the patient to die at home. Progressively, as patients lost their autonomy (in particular if not given the necessary support network), there was a question of bringing patients back to the hospital to die. The following is the story of a patient, but also a fairly common situation discussed in these meetings:

> An older man (60 years) is at the end of life due to a rapidly progressing cancer. He wishes to die at home; however, the situation has become impossible. The healthcare team meets to discuss what to do. Unfortunately, he is unable to ensure his own self-care, with worsening hygiene problems impacting his health and making his house unsafe. While they would have preferred to respect his wishes to die at home, as the man lives in a rural area, a nurse cannot visit regularly, and he does not have a family around to support him. What can the team do for the patient? Somehow there are no good solutions to be found. His worsening cancer is leading to psychiatric issues, yet psychiatry considers him an oncology patient. He is too young for a retirement home as well. They wish to help him but do not know what to do other than putting him under guardianship and hospitalizing him, even if it may be against his wishes.

In this type of situation, it would be surprising if these healthcare providers—faced with difficult (or impossible) decisions in how to do the best for their patient—were not emotionally impacted. They face tough decisions in which there was no "right" answer and they had to come to terms with limits about what they—and their institutions—could realistically do to help their patients. They faced anger at the injustice of the situation, sadness about the upcoming death of their patient, frustration at patients and their families who were at times angry toward them despite their efforts to help them, and anger with themselves about the limits of the institution and of medicine to help them, particularly in cases like these, where vulnerable patients fall between the cracks.

In these situations, affects surfaced in healthcare providers' narratives of their patients, including in their ways of describing them and their conversations with them, but also in their non-verbal language when they talked about patients with which they were particularly close or in conflict. It came out in their ways of describing small details of their personal lives or situations they had witnessed, such as the lack of hygiene in the home, or the person's social isolation. These discussions were a safe space for healthcare providers' affects to be recognized, as a well as a recognition of those lived by patients and their families. Because of this, these affects became a part of the collective narrative—and therefore the decisions taken. However, that does not mean healthcare providers' affects were formally part of the healthcare plan or the narrative (decision) written afterward in the patient file. They were given a certain space in the collective, but with some limits: they could not go too far, as they did not follow the standard rules of reporting.

In addition, the affective neutrality demanded for in interdisciplinary work was also witnessed by the researchers in these spaces. Should a certain affect dominate the discussion (such as anger toward the patient or the institution), or if they were seen as inappropriate or unhelpful to the discussion, healthcare providers did not encourage their expression. Healthcare providers interviewed stated their affects were not necessarily in the "right" place during these meetings (which was geared toward the patient). Thus, they did not hesitate to stop them or steer the conversation away from them. However, this did not mean they were ignored. Healthcare providers interviewed saw them as a sort of "red herring," alerting to the need to address difficulties in the team or in the hospital. In this way, they served as a resource about care and team relationships: affects were taken seriously, in particular negative ones, and if needed, "retold" to hierarchies and management.

In light of this discussion, we can see such types of team meetings can at times—but not always—provide a safe space for affects to be expressed and help work toward person-centered care, as it helped the team see a wider perspective on their patient and their relationships. We also see the possibility that healthcare provider's affects, in functioning as "red herrings," be heard at a service or institutional level. The question however then becomes how managers/heads of team will understand them as saying something important about care.

Having explored the possibilities of these interdisciplinary meetings, let us now turn to our second example, namely analysis of practice for young doctors. These meetings, as we will see, also provide a safe space for collective affective sharing, although with the limitation of having them be shared among doctors who are not necessarily in the same service. The idea of analysis of practice is for each professional to bring a difficult case to a group of professionals so they can discuss what went right/wrong, including how the doctor lived through the experience. The goal is to anticipate future cases and to benefit from collegial reflection. The discussions are often facilitated by a senior doctor; however, a psychologist may also be present to focus on the psychological impact of such cases. In their testimony, young doctors are able to freely express their affects, including fear, in a non-judgmental environment, with other persons facing similar difficulties, as professionals are still in the phrase of "learning" and are encouraged to speak openly on what they experienced and how they lived it. Providing a closed group for young doctors helps create an environment in which affects are not judged. The problem with this method is the affective neutrality already engrained from medical school (see Chapter 8) may also happen in these spaces, unless the facilitator actively encourages its expression. Here the presence of the psychologist is in particular adapted to this task and may help young doctors to express their emotions with less restraint.

From these two examples, there appears to be a feasible place for affects to be told but also to be recognized by others through collective discussion. If done carefully, such methods can help healthcare professionals see and recognize them as important to care practice. As we can see, this can happen both in interdisciplinary reflection, when they are able to witness that their colleagues are also emotionally affected by the situation (and being emotionally affected is no bad thing, as it related to how they care for their patient), but it can also function in groups geared toward specific professionals and their affective needs, as we saw in the example of analysis of practice.

4.6 Conclusion: Opportunities and Limits of Narrative Methods

The four methods proposed in this chapter have the potential to serve as affective resources, opportunities, and forms of recognition, thereby helping us move toward greater affective injustice for healthcare providers and patients. Table 2 below summarizes the main points, including their possibilities and limits.

Method	Narrative method	Level	Opportunities/ limitations	Limitations
Free-form journals	Written	Individual	Free space for affective expression without restraint, including expression of negative emotions	Passive role of writers in use/ transmission of knowledge; little space for institutional learning led by healthcare teams
Narrative medicine workshops	Written, oral	Individual—small group	Free space for expression; affective sharing in small groups; may create a change in perspective if invested in and encouraged long-term	Narratives not necessarily shared outside of a small group; few possibilities for institutional learning
Narrative banks	Written, oral, drawings	Individual—hospital - management	Free space for expression; can permit higher level learning at the strategic level and affective sharing in the hospital	Passive role of writers in use/ transmission of knowledge; no space for affective sharing in small groups

(continued)

(continued)

Method	Narrative method	Level	Opportunities/ limitations	Limitations
Interdisciplinary reflection	Oral	Hospital service—management	Somewhat safe space for individual and collective expression, particularly to improve solidarity among the team; affects may also alert to problems and crises in the service/hospital	Forum not always adapted and/or does not always enable healthcare providers' affective needs to be addressed; possibility to improve reflection at the strategic level is limited and based upon how hierarchies recognize affects

As seen from the above table, there is no one narrative method that can fully provide for the two ambitions of facilitating: (1) affective resources and opportunities of healthcare workers to foster their well-being and provide a safe space for emotional expression; (2) being able to use these affective resources and opportunities to improve healthcare delivery. For instance, while the narrative bank can, if done in appropriate ways, help affects enter into institutional learning, it does little to provide affective opportunities for the individual in their daily healthcare work or to value these affects in a team, as they remain anonymous. Similarly, while interdisciplinary reflection may help value affects in teams, it is not always welcomed as it may not be seen by all as the appropriate place for their expression. Thus no one narrative method can fully satisfy both ambitions in this chapter; however, taken together, they do accomplish these goals. This suggests healthcare institutions should invest in multiple forms of narrative methods to work toward affective justice in care institutions.

Before concluding, however, let us be realistic about the use of narrative methods in hospitals. While many healthcare professionals may wish to practice narrative-based methods in their work, the lack of time, increased patient loads, as well as the increasing use of technical tools means that in practice, patient narratives may be relegated to the background. If an attention to narratives has re-emerged, it is also because of these challenges of paying full attention to the patient's narrative. Taking this into consideration, if we are use narrative methods as a way to value affects in healthcare institutions, we will have to deal with the marginal place they currently have in healthcare practice and find convincing ways to advocate for their inclusion. In particular, there will be a need to invest in intervention research to demonstrate how narrative methods: (1) improve well-being of healthcare workers; (2) help improve healthcare provision, in particular by providing a greater diversity of experiences/perspectives about care; (3) contribute to institutional knowledge, in particular when integrated into strategic-level discussions on care provision (narrative banks), or altering to dysfunctions in the team (interdisciplinary reflection). By showing their pertinence to care provision, they are more likely to take a sustainable place in healthcare institutions. In addition, it should be shown that these initiatives are not necessarily expensive. Initiatives such as free-form journals, narrative banks, or narrative method workshops can be organized at a minimal cost; however, in order to be sustainable, they will need to be supported by the institution (through worktime), as the narrative medicine workshop showed.

Before concluding, another point of vigilance is necessary, namely, the risk of taking the affects expressed through narration fully at face value. Indeed, what we can learn from the narrative bank and the free-form journals is that narratives, when written for an audience, are not fully free or disinterested. When writers know they will be read, we can expect they will be influenced in how they narrate their stories. This includes what emotional impact they are seeking from the reader in their telling. We can for instance question in the example of the free-form journals whether negative effects (anger and disgust) were used to push their specific agendas, such as problems with hierarchies. Similarly, in the case of the narrative bank, healthcare providers who knew their testimonies might be read at a higher level might feel pressured to focus on more positive aspects of care. This is the risk of taking affects at face value to guide institutional change, particularly when the writer does not engage actively with the reader. We will therefore need to remain cautious of those methods where writers are disengaged from collectively sharing.

To conclude our discussion, we have seen that narrative methods represent a promising method to work toward greater affective justice for healthcare providers (and in some ways for patients) in the hospital. However, they are not enough. In order to deal with the challenges of each method, we will need to invest in several forms of narrative methods, as well as consider methods that make the writer more responsible for their testimonies so they can become, as Ricoeur has told us, capable beings in the stories of their lives. We will also need to pay attention to the goal of the methods we have chosen and to be realistic about our ambitions. For instance, is it to provide an individual safe space for affects (free-form journals), provide affective sharing spaces (narrative workshops), promote team building and engagement (reflection on cases), or provide a research and knowledge base to the hospital (narrative banks)? These ambitions will guide our choice of methods.

This discussion also shows we will need to invest in other methods to create an affective culture change at the hospital. In the next chapter, I will explore the role of spirituality—and how taking seriously both patient's and healthcare provider's spiritual needs—may provide also a safer space for affects. In later chapters, I will investigate how patient partners/experts can help move toward this goal, as well as the role of healthcare design and architecture in providing affective resources.

References

Bajaj, Nimisha, James Phelan, Erin E. McConnell, and Suzanne M. Reed. 2023. "A Narrative Medicine Intervention in Pediatric Residents Led to Sustained Improvements in Resident Well-Being." *Annals of Medicine* 55 (1): 849–859. https://doi.org/10.1080/07853890.2023.2185674

Bogaert, Brenda, and Jean-Philippe Pierron. 2022. "From Moral Distress to Mutual Recognition: Diaries Kept by French Care Professionals During the Covid Crisis." *Ethics and Social Welfare*, July, 1–16. https://doi.org/10.1080/17496535.2022.2089710

Bourquin, Céline, Amaelle Gavin, and Friedrich Stiefel. 2022. "La Liaison Psychiatrique I : Clinique." *Revue Médicale Suisse* 18 (769): 261–64. https://doi.org/10.53738/REVMED.2022.18.769.261.

Charon, Rita. 2008. *Narrative Medicine: Honoring the Stories of Illness*. 1. paperback ed. Oxford: Oxford Univ. Press.

CHUV. 2024a. "Le projet PENbank." 2024. https://www.chuv.ch/fr/psychiatrie/dp-home/recherche/centres-et-unites-de-recherche/groupe-de-recherche-du-service-de-psychiatrie-de-liaison-pli/le-projet-penbank

CHUV. 2024b. "PenBank: récit 16." *Infine*, January 19, 2024.

Delorenzo, Christian, Tsellina Desfemmes, Marina Vignot, Jean-Marc Baleyte, and Rita Charon. 2021. "Des Groupes de Médecine Narrative Dans Un Centre Hospitalier : L'expérience et Le Dispositif Du Centre Hospitalier Intercommunal de Créteil (CHIC)." *Revue Médecine et Philosophie*, December 31, 2021, 5 edition. https://medecine-philosophie.com/wp-content/uploads/2021/12/12_Delorenzo_Baleyte__6.pdf

Ekman, Inger, Zahra Ebrahimi, and Patricia Olaya Contreras. 2021. "Person-Centred Care: Looking Back, Looking Forward." *European Journal of Cardiovascular Nursing* 20 (2): 93–95. https://doi.org/10.1093/eurjcn/zvaa025

Gallegos, Francisco. 2021. "Affective Injustice and Fundamental Affective Goods". *Journal of Social Philosophy* n/a (n/a). https://doi.org/10.1111/josp.12428

Kalitzkus, Vera, and Peter F Matthiessen. 2009. "Narrative-Based Medicine: Potential, Pitfalls, and Practice." *The Permanente Journal* 13 (1): 80–86. https://doi.org/10.7812/TPP/09.996

Keen, Suzanne. 2015. "Narrative Emotions". In *Narrative Form*, by Suzanne Keen, 152–161. London: Palgrave Macmillan UK. https://doi.org/10.1057/9781137439598_12

Kleinman, Arthur. 1989. *The Illness Narratives: Suffering, Healing, and the Human Condition*. New York: Basic Books.

Reed, Pamela G., and Cindy J. Rishel. 2015. "Epistemic Injustice and Nurse Moral Distress: Perspective for Policy Development." *Nursing Science Quarterly* 28 (3): 241–244. https://doi.org/10.1177/0894318415585634

Remein, Christy DiFrances, Ellen Childs, John Carlo Pasco, Ludovic Trinquart, David B. Flynn, Sarah L. Wingerter, Robina M. Bhasin, Lindsay B. Demers, and Emelia J. Benjamin. 2020. "Content and Outcomes of Narrative Medicine Programmes: A Systematic Review of the Literature through 2019". *BMJ Open* 10 (1): e031568. https://doi.org/10.1136/bmjopen-2019-031568

Ricœur, Paul. 2006. *Parcours de la reconnaissance: trois études*. Les essais. Paris: Stock.

Ricœur, Paul, and Kathleen McLaughlin. 2009. *Time and Narrative. Vol. 1*. Repr. Vol. 1. Chicago, Ill.: Univ. of Chicago Press.

Sacks, Oliver. 1998. *The Man Who Mistook His Wife for a Hat and Other Clinical Tales*. 1st Touchstone ed. New York, NY: Simon & Schuster

Chapter 5
Affective Justice and Spiritual Care

Abstract This chapter will discuss the affective resources, opportunities, freedoms, and recognition offered by spiritual care. While practices and rituals are diverse, they include an important affective dimension in their methodology and practice, from helping patients process their emotions at the end of life to supporting healthcare providers to work through their moral distress in difficult cases. In this chapter, I will first of all introduce the concept of spiritual care and the link between spiritual care and affective justice. I will then introduce several methods of spiritual care—in particular spiritual history taking and the role of the chaplain—to better understand in what ways care may be delivered. To better understand its articulation in practice, I will also consider two instances in which spiritual care may be particularly relevant for patients (palliative care and to better address the affective needs and values of marginalized populations). I will then consider the importance of spiritual care for healthcare providers, notably by discussing the role of the hospital chaplain during the pandemic. Throughout this journey, I will show the interlinkage of spiritual care, affects, and value-based care for healthcare providers, patients, and their families. I will also discuss the limits of delegating emotional work to spiritual care professionals.

Keywords Spiritual care · Emotional distress · Affects · Hospital chaplain · Spiritual beliefs and practices

5.1 Introduction

The word spirituality is derived from the latin "spritus," meaning breath, inspiration, character or soul (Lalani and Chen 2021). While spirituality and religion are often used simultaneously, most researchers agree that these terms are separate constructs and that spirituality is a broader concept than religion. Scholars notably make the distinction between religion, associated with an organized system of beliefs, practices, and symbols, and spirituality, which involves a person's search for meaning or purpose in life, and may involve but also go beyond an organized religious system.

B. Bogaert, *Affective Injustice in Healthcare*, The International Library of Bioethics 112, https://doi.org/10.1007/978-3-031-94378-2_5

Spirituality is believed to be person-specific and context dependent, connected to the values and beliefs of the individual (Lalani and Chen 2021). Because of this, individuals may engage in spiritual practices in other ways than in organized religious rituals, including through arts-based practices such as art, music, or drama.

In this chapter, I will elaborate the possibilities of the growing movement known as spiritual care. This movement refers to the provision of healthcare that addresses the person's existential and spiritual needs in connection with their illness (Hvidt et al. 2020). Studies show that many patients consider spirituality important in their lives and an important part of their illness experience (Hall and Powell 2021). While there is no consensus on what exactly spiritual care is and who should be involved, it may involve having a chaplain accompanying the patient at the bedside, but it can also start at the beginning of a healthcare journey with the doctor or nurse taking a spiritual history. This means that spiritual care may include but also go beyond the figure of the hospital chaplain and implicate a diversity of healthcare providers who can support the patient with spiritual beliefs—whether or not they have these beliefs and practices themselves. As we will see, by involving a diversity of healthcare actors, spiritual care can be integrated into a person-centered care perspective that involves recognizing and providing care in line with the person's values.

To better understand the movement of spiritual care, let us take a movement to better understand its formulation in healthcare institutions today. First of all, while the name "spiritual care" may be relatively new, it is of course not a recent phenomenon, just like the narrative movement discussed in the previous chapter. Theories of health as a harmony between body, mind, and spirit have guided the philosophy and practice of medicine since Antiquity. Healthcare professional education and training have also been historically linked with spiritual organizations, at least in some professions. During the early twentieth century, most nursing schools were affiliated with religious institutions and were trained to provide a care that focused on ideas of harmony between mind, body, and spirit (Lalani and Chen 2021). While such concepts moved to the background in the shift toward evidence-based practices, the limits of these models in recent years and the promotion of the patient/person-centered healthcare movements have brought back these debates.

Spiritual care has emerged in a context in which we are increasingly criticizing the limits of "modern medicine." However, in a context of rising costs for healthcare and the need to focus on patient outcomes, it is also advocated from a public health perspective. With these perspectives in mind, research is increasingly showing the benefits of spiritual care in terms of health outcomes. Research has shown for instance that for those with a positive relationship toward their spiritual beliefs and practices, such ideas can provide coping resources, protect against depression, and enhance pain management (Larson and Larson 2003). In turn, failure to address the patient's spiritual needs has been associated with poor quality of life, dissatisfaction with care, and increased costs at end of life (Astrow et al. 2007; Koenig et al. 2017). There is therefore a movement to promote spiritual care in healthcare practices in order to deal with the limits of evidence-based medicine, but also because it has shown promising results from a public health perspective.

To be clear, I am not advocating healthcare institutions become religious or spiritual-centered, but rather, in a person-centered perspective, they give greater attention to those patients (and healthcare providers) whose spiritual beliefs are important to them and how they envisage their illness. In other words, I am seeking to provide resources and opportunities in line with their values, and these values include spiritual beliefs and practices. Furthermore, relevant to the debate on affective injustice, we will see that affects are an important component of how spiritual care is delivered, conceived, and received, and has a relationship with the positive healthcare outcomes demonstrated by research. Taking the diverse possibilities of spiritual beliefs and practices in mind, this chapter will focus on the potential of spiritual care as a way of working toward the provision of an affective safe space for patients, families, and healthcare providers.

However, before beginning our discussion, it is important to highlight its contours and limits. In the first place, this discussion will largely center on spiritual care in the Western world, where spiritual beliefs and practices have an uncertain place. In many countries around the world, spirituality has never been sidelined in theories of health, well-being, and healthcare practice. It is therefore a strategic choice on my part to focus on the Western world, where such practices have been marginalized in the face of increasing technicity and the promotion of evidence-based medicine.

Furthermore, most of the research I will discuss is based upon Christian-based practices, which have largely dominated research on spiritual care (although this is changing). However, it is important to note that spiritual care provides the kind of affective resources and opportunities we are looking for regardless of a specific religious or spiritual tradition. It will also be important to recognize the heterogeneity among Western countries in how spiritual care is conceived and practiced, depending on cultural and legal issues surrounding the place of religion in society. Indeed, if we compare two countries in relative proximity to each other, Switzerland and France, we can see large differences in how spiritual care may be received and recognized. In Switzerland, spiritual care is emerging as an important research area and is of interest to healthcare institutions for its potential to provide more value-based care. In comparison, in France, spiritual care remains taboo and underdeveloped, as one of the founding principles of the country is its secularity. These contextual factors are important when considering how to find a realistic place for spiritual care—and thus to work toward affective justice through such methods.

Having elaborated these limitations, let us now turn to a second introductory discussion on the link between affects and spirituality to better understand in what ways the integration of spiritual practices may help work toward affective justice in healthcare.

5.2 Affective Justice and the Role for Spirituality

In this section, I will propose a rapprochement between the concept of affective injustice and spirituality that will guide us throughout this reflection and the case studies. As stated in the introduction to this book, I am interested in those practices that seek to increase, bring out into the open, and provide a safe space for the affects of healthcare providers, patients, and families. I believe that spiritual care may help work toward this goal by responding to the patient's and family's affective needs, including discussing those affects they feel and that may help them apprehend their illness journey or in supporting someone with an illness. I will advocate that spiritual care may benefit both those with spiritual beliefs and practices, but also build an "alternative" means of care that may help provide a safer affective space even for non-believers. Furthermore, I will show how, even though spiritual care may be targeted toward patients and families, it may also help healthcare providers. In line with Gallegos' (2021) proposal that has been used throughout Part II, I will show the possibilities and limits for spiritual care to facilitate a number of affective goods for patients, families, and healthcare providers. These include:

- Affective resources and opportunities: providing materials, activities, and circumstances for self-care for patients, their families, and healthcare providers
- Affective freedoms: respecting the person's values, including freedom from interference in the pursuit of subjective well-being
- Affective recognition: considering and responding to the person's affective needs

To better understand how spiritual care may help with these ambitions, I will need to turn to studies in psychology, which in recent years have helped to make the link between affects and spiritual practices. I will first of all discuss how spirituality can have a positive effect on the health and well-being of patients with spiritual beliefs, which has already shown its value from a public health and person-centered perspective.

In psychology, it has been recognized that positive affects—including joy, gratitude, and hope, among others—broaden one's mindset and enable a person to build enduring psychological, biological, and social resources for survival (Van Cappellen et al. 2023). Research studies in psychology and spirituality are increasingly making the link between positive emotions and spiritual beliefs and practices (Koenig 2012). While these studies have largely been based upon Western, Christian traditions, it has also been shown that the link between spirituality and positive emotions hold across different religions and spiritual beliefs and practices (Van Cappelen et al. 2023).

How does this happen in practice? To understand these mechanisms, it is necessary to turn to the emotional regulation strategies employed by the person in function of their beliefs. Spirituality shapes the person's beliefs about emotions and those they should experience. This leads to several types of emotional regulation strategies: *situational acceptance* (accepting a given situation such as an illness and adjusting one's mindset in function) and *cognitive reappraisal*, involving reframing an emotional event by changing its perception and meaning (Vishkin et al. 2019). These adaptive strategies, which help the person to accept their situation and adapt their perception, explains why persons with spiritual beliefs and practices often experience positive emotions such as hope in illness. These kind of adaptative strategies are intrinsic emotional regulation strategies that may help the person not to "give up" by keeping a positive outlook.

Another way to explain how spiritual beliefs and practices may influence the person's subjective well-being, and why it will be important to provide spaces for affective freedoms through spiritual care, has been demonstrated through the concept of relational spirituality (Augustyn et al. 2017; Hall 2004). According to Augustyn et al. (2017), we can understand the person's relationships to a spiritual figure in association with the emotional attachments that infants form in their first years. By extending these child development attachment theories to spiritual experiences, spiritual beliefs and practices may replace the caregiver as the attachment figure. By providing feelings of secure attachment, this may permit the person to feel supported and reduce their isolation.

These studies from psychology show that the affects surrounding spiritual beliefs are not necessarily restrictive or limiting. It may encourage them to exercise greater agency, as emotions such as hope may encourage them to seek out other resources to support them—whether it be patient or spiritual organizations (Kim-Prieto et al. 2014). Spiritual beliefs and practices in this way may be compatible with having the person take a greater role in their healthcare and help reduce their isolation. In turn, while these spiritual beliefs and practices could well stay in the private domain, when they become recognized and cultivated as part of a person-centered care plan, it has a number of advantages. First of all, healthcare providers can help those with spiritual beliefs and practices to identify outside resources, such as patient groups with likeminded spiritual beliefs to help support them or find information about the illness. It can also foster greater dialogue and shared decision-making with the healthcare provider, particularly when they take into consideration the patient's spiritual beliefs in the treatment plan. Given these perspectives, it does appear that working with patients in terms of their spiritual needs may help work toward greater affective justice, in particular when the hospital space provides affective resources, opportunities, freedoms, and recognition through spiritual care.

However, as we will explore in the next section, spiritual care is not necessarily limited to those with spiritual care and concerns. Indeed, it also serves as another, different means of practicing and reporting on care. While I will discuss some promising methods in the next session, let me take the example of a discussion session lead by a hospital chaplain in which everyone—regardless of their spiritual beliefs or non-beliefs, is invited to attend. There is a common misperception that the role of the hospital chaplain is to prophesize their own specific religious beliefs; however, in practice, hospital chaplains incorporate a diversity of religious beliefs and practices into their practices. They are therefore accessible and relevant to a diversity of religious beliefs and non-beliefs. If we take for example a discussion session on a difficult case led by a hospital chaplain, this has several advantages from the affective justice perspective. First of all, as a non-medical professional, there may be less expectations about them to be "affectively neutral." This means that they may provide a safe space in which positive emotions (such as hope) but also negative emotions (fear, anxiety) may be expressed and recognized. Secondly, by bringing together several persons, they can provide a collective affective sharing space in which affects are discussed and taken seriously. Whether or not these affective concerns may be tied to existential or spiritual concerns will depend on the group and their needs. As will be discussed throughout this chapter, such methods can provide both a safe affective resource for healthcare providers, but also promote a wider safe affective space, where affects are given place and recognition in sharing about care experiences. It is also not to be forgotten that healthcare providers themselves have their own spiritual beliefs and practices, and the difficult care situations they face will inevitably influence how they experience and perceive these situations. This means providing for the spiritual care of healthcare providers is not to be forgotten and is a way of taking care of them, as well.

5.3 Methods to Integrate Spiritual Care into Healthcare Practice

Having established the link between spirituality and affects, it will now be important to understand the tools that will permit this to be formulated and integrated into care practice. As a first step, it will be important to acknowledge this is a nascent field. Indeed, despite a significant number of patients (41–94% depending on the study) who express the desire for clinicians to address their spiritual concerns (Borneman et al. 2010), spiritual care remains an underdeveloped field. While the reasons for this are complex, studies in particular have highlighted the healthcare provider's discomfort in discussing spirituality with patients and their lack of training to provide spiritual care (Koenig et al. 2017), as well as inadequate support from healthcare institutions. In addition, there are practical challenges to take into account. While there is increasingly advocacy for spirituality care, it is not always easy to put into practice as each person may have unique practices and beliefs. This section will highlight some promising methods that promise flexibility across the diversity of religious traditions. The section on case studies will also highlight how to adapt them for the specific religious/spiritual tradition.

Spiritual history: a first step to integrate spiritual care into healthcare practice involves taking a spiritual history. Its purpose is to learn about how patients cope with their illness, the support systems available to them, and to identify any strongly held beliefs or practices that may influence their medical care (Lucchetti et al. 2013). Studies show a high number (up to two thirds in one study) believed their doctors should be aware of their religious or spiritual beliefs (MacLean et al. 2003). However, despite of the importance to patients, studies indicated only a small minority (around 10%) of healthcare providers currently do spiritual histories (Koenig et al. 2017). The reasons for its low implementation have already been discussed: lack of training and discomfort of the healthcare provider in discussing spiritual needs with patients.

Despite this low implementation, a number of tools have been developed to give providers a clear method for the spiritual history of patients, including FICA, SPIRITual History, FAITH, HOPE, among others (Lucchetti et al. 2013). In this section, we will briefly look at the FICA tool, as it is the resource the most discussed, researched, used and practiced. The FICA (Faith, Importance, Community, Address) Spirituality History Tool was created by Dr. Puchalski in collaboration with primary care physicians in 1996. It provides a way for the clinician to integrate open-ended questions into a standard medical history. It is based upon four domains of spiritual assessment (Puchalski and Romer 2000):

1. **F**aith or beliefs of the person
2. **I**mportance of spirituality in a person's life and the influence on the person's healthcare decision-making
3. The person's spiritual **C**ommunity
4. **A**ddressing spiritual needs

This tool is a simple but effective means of working with the patient (should they desire) to discuss their faith and beliefs, how it may influence their health and decision-making, and to understand the extent of their support network. While it is an explicit method that may make some patients uncomfortable if they do not wish to discuss their spiritual beliefs and practices with their healthcare provider, it takes the lead from the patient and what is important to them. Therefore, for those who wish to keep these beliefs and practices in their private domain, it can be stopped or adapted to their needs.

The fourth step of the tool provides the way in which spiritual care can be delivered by enabling a discussion with the patient on the ways the healthcare provider can support them in their spiritual needs, or help them to identify an appropriate referral person (psychologist, member of faith community, hospital chaplain, etc.) to address these needs. In this way, tools such as FICA can help work toward a person-centered perspective, including identifying how the person's spiritual beliefs may affect their healthcare decision-making, but also permitting a better understanding of the community and health resources important to their healthcare plan. Tools such as FICA do not necessarily increase history taking in a consultation, but can be integrated into the more global history taken by the clinician of their patient.

Another benefit is that it may also alert the healthcare provider to the more negative effects of spiritual beliefs. Indeed, while I have so far discussed the many positive effects of spiritual beliefs and practices, it is necessary to take into account that not all persons will necessarily feel positive emotions surrounding beliefs in a higher power. In a study by Beck (2006), it was shown those who developed a positive relationship to their spiritual beliefs and practices experienced this power as more loving and less controlling (thus associating positive emotions to their spiritual beliefs and practices), whereas those with more insecure attachment had a more anxious relationship. By taking a spiritual history, the healthcare provider may pay attention to these individual perceptions to better understand how their spiritual beliefs and practices may be also be a constraint in their illness.

Furthermore, while tools such as FICA have mainly been developed for clinicians, this does not mean other healthcare providers cannot use them. While in the context of general medicine, the doctor can use FICA as an important resource in getting to know their patients, in the hospital context other professionals may be mobilized for this kind of history taking. This has been suggested in particular for nursing. Given the closer relationship of nurses to patients in the hospital context, and the fact that nurses' theories and education are increasingly integrating spirituality into standards of practice, nurses may also take on this responsibility. However, other authors (Selman et al. 2018) show that all staff caring for patients should be trained to be able to take a patient's spiritual history, as it not only promotes a person-centered care perspective, which will help improve the quality of care, but may also create a space for affects to be taken seriously and valued. A patient may feel more comfortable discussing their spiritual beliefs (and associated affects about these beliefs) with someone of the same faith, or a member of the care team with whom they can identify or feel more comfortable with, rather than an appointed contact person for taking spiritual histories. By promoting a culture of attention to spiritual care in institutions, whereby all healthcare providers will feel trained and comfortable in addressing the spiritual needs of patients, we can work toward an environment in which the person's spiritual beliefs and affects are taken more seriously.

Hospital chaplain: another common method in spiritual care is the integration of a trained chaplain to support the patient's spiritual needs (Larson and Larson 2003). The benefit of a hospital chaplain is to give a trained resource person whom the patient and healthcare team can turn to for their spiritual and affective needs. At the current time, while the figure of a hospital chaplain is the most visible form of spiritual care, research has shown that the majority of consults by a chaplain occurs in the final day of the patient's life. This temporality limits their ability to meet the spiritual needs of non-terminal patients or those facing terminal diagnoses in early stages, in particular as it has been shown that the intervention of a chaplain is associated with improvements in the patient's quality of life (Kirchoff et al. 2021). From this perspective, they should not be limited to the final days in the person's life and can provide more holistic care in the hospital if they are integrated into other services.

Furthermore, the support that hospital chaplains provide is not limited to the spiritual needs of patients. It can notably include their affective needs. By promoting a one-on-one or group discussion with patients about their experiences of care and illness, affects become an important way of telling and understanding illness. It can therefore provide the kind of safe space and affective resources and recognition we have been seeking throughout this book. In addition, having a key resource person integrated into a care team but not directly involved in the diagnosis and treatment enables someone who can focus on the more existential and affective needs not always given adequate space in the care plan.

In addition to supporting the patient's spiritual needs, hospital chaplains can also provide the resources and opportunities for affective justice for healthcare providers, including helping to relieve their feelings of moral distress. The possibility of having this key resource person take into account healthcare providers' needs has been increasingly advocated for in the post-COVID period, as chaplains played an important role in helping healthcare providers deal with difficult decisions during this period. While the methods used by hospital chaplains are diverse, among effective methods include forms of ritual to help healthcare providers express or reframe feelings and experiences. This can take the form of "debriefings" with hospital staff after the difficult death of a patient. During these debriefings, a hospital chaplain can facilitate a discussion with healthcare staff on their physical, affective, and spiritual experiences. For instance, an affect such as gratitude, sadness, or hope may be the theme for a discussion and can help healthcare providers voice their experiences, thereby helping to relieve their moral distress. As Klitzman et al. 2022) have shown, while rituals may encompass a variety of goals and functions, they often include expressing and exploring emotions and providing support to reframe events and feelings positively or to work through difficult emotions and experiences. For instance, in Klitzman et al.'s study (2022), one chaplain describes his method as, *"we gather everybody involved and give space for processing the emotions, like a root cause analysis at a Mortality and Morbidity conference, but dealing with emotional, rather than strictly medical sides of the decisions made. When people try to second guess the clinical decisions, we redirect them toward, 'there'll be a context for that, but what was it like for you? What were you feeling when you were making that decision?"* As seen from this quotation, it is not a question of steering healthcare providers from more negative emotions to more positive ones, but rather helping them to reflect on the emotion and existential aspects of difficult cases, thereby facilitating a safe affective space in which their affects are taken seriously and recognized. This also includes taking seriously negative emotions such as disgust or anger for what they represent. This example shows that spiritual care sessions function in a similar way to the narrative-based methods explored in the previous chapter. While the goal of these methods is not necessarily affects per say, in the telling of care experiences, affects naturally take a center stage and are often the means in which these experiences are related and understood. By promoting such free spaces of expression, we are therefore promoting new ways of care toward patients and healthcare providers, as well as providing a safe space for affects.

However, despite these promising indications on spiritual care as a way to provide for the affective needs of healthcare providers, it is to be recognized that these kinds of support currently suffer from institutional resistance and lack of support, limiting the capabilities of hospital chaplains to provide spiritual care to this group. According to Klitzman et al.'s study (2022), which interviewed hospital chaplains about their methods and experiences, one of the findings was that many hospital administrators resist, minimize or dismiss staff's spiritual needs. Indeed, this study is a good example of the extent of affective injustice suffered by healthcare providers from their institutions. The study notably showed instances in which chaplains were told by hospital management that there was no space for such emotions about difficult care experiences and they should just "suck it up." While this should give pause, there were a number of ways the study showed how chaplains could overcome such resistance. A facilitating factor was to help management gain trust in the individual person, which often included the time spent in the service and their interactions with healthcare providers. There was also the need to clarify that the service they provide was not necessarily religious, but rather adapted to each person's needs, including their affective needs. In this way, their presence became more accepted in the service.

Spiritual care theories and affects: in this final section, it will be necessary to pay attention to the diversity of ways in which spirituality is practiced and theorized[1] in terms of affects. There are at least four models of spiritual care currently in practice and their relation to affects is different. The first type of spiritual care is closely related to the nursing sciences. This type of spiritual care—in line with nursing theories—seek to value time, listening, and compassion. In this theory, spiritual care professionals see affects as a central part of their healthcare delivery, and they will actively seek to discuss and bring out the patient's emotions in order to provide person-centered care. Therefore, this first model directly integrates affects and makes them a priority in the healthcare plan. Spiritual care in this way functions as a meeting place, in which emotions may be related to their healthcare plan, but also other factors in their life. The goal of the chaplain will be to provide this affective resource and opportunity for the patient.

The second type of spiritual care stems from explicit religious requests by patients toward their hospital chaplain, such as prayers and other rituals to help the person face death. Here again, affects are important to the planning of the ritual, and in the request by the patient, such as "fear of death." Requests are also explicitly formulated by the patient toward the healthcare provider, such as hope and prayer for the "wisdom of the healthcare provider" in their treatment. In this model, too, affects are explicitly linked to the act of spiritual care.

The third type of spiritual care looks for relevant information to guide spiritual care practice. In the model used by the university hospital in Lausanne, for instance, four concepts are taken into consideration in spiritual care: (1) meaning; (2) transcendence; (3) values; (4) psycho-social identity (Monod et al. 2010). In the first theme, "meaning," affects form an important means of understanding and dialogue with the patient. They serve as resources to know better their patient and to work with them. For instance, when a patient expresses anger, the spiritual care practioner will discuss with the patient why they are angry. This means taking the anger seriously as a source of knowledge. They will then use this emotion as a guide to help them think about their illness in its spiritual and existential dimensions. The spiritual care practioner in this way accompanies the patient toward those emotional regulation strategies (such as cognitive reappraisal) that may be beneficial for them to apprehend their illness.

The fourth model of spiritual care is the only one ambiguous about the value of affects. While it recognizes the importance of affects to create a therapeutic alliance between the spiritual care provider and the patient, it is cautious of having emotions be the guide of spiritual care practice. This is because not all persons experience emotions in the face of bad news, and those who do so may not necessarily have a

---

[1] I would like to thank Etienne Rochat, a hospital chaplain at the Lausanne University Hospital and researcher in spiritual care at the Institute for Humanities in Medicine in Lausanne, Switzerland, who provided these distinctions in the various models of spiritual care and the place for affects in each of them.

deep experience or relationship between their affects and their existential experience. However, we can see that this last example is also turned toward person-centered care, designed for the needs of the patient, and this includes their affective needs, when relevant to the individual.

5.4 Case Studies of Spiritual Care and Affective Justice

From the four models described above, we can see spiritual care practices pay particular attention to affects (when relevant) and they form an integral part of how they provide person-centered care. However, to better understand how spiritual care can improve care experiences, particularly in difficult to treat or stigmatized illnesses, it will now be necessary to go in the field to understand how spiritual care happens in practice.

I have chosen three case studies to elaborate this subject. The first is palliative care, where the role of spirituality to accompany patients at the end of life remains the most developed form of spiritual care. In the second, I will return to the problem from Chapter 3, namely provision of mental health care for marginalized groups. In the third, I will look at healthcare providers' needs and the role of spiritual care during the COVID pandemic. Therefore, while the emphasis in this section will be to how spirituality provides a safe affective space for patients and their families, I will also not forget the affective needs of healthcare providers and how they may benefit from this alternative means of caring.

5.4.1 Palliative Care: An Important Role for Affects

Palliative care is a natural place to start the discussion on the place for affects in spiritual care practices. It is both the place persons naturally think of when they think about spiritual care (given the important existential questions facing patients at end of life), but also the one in which there has been the most advocacy for its inclusion. However, as the hospital chaplain's role is (almost) evident in this service, I have decided to focus here on the role of other care providers, and in particular the nurses' role. In this section, I will also be interested in whether explicit or implicit strategies for spiritual care should be promoted.

However, the first question to ask perhaps is why should we be interested in this issue? The answer is spiritual care is important for patients. Research shows patients wish to have access to spiritual care at the end of life. They show that desire for spiritual care increases strongly with the severity of the illness, with nearly 50% wanting spiritual care when faced with near death (MacLean et al. 2003). However, despite its importance to patients, spiritual care remains the least developed and neglected dimension of palliative care, and training in spiritual care for palliative care personnel is limited (Gijsberts et al. 2019).

Palliative care is defined as, "the active holistic care of individuals across all ages with serious health-related suffering due to severe illness and especially of those near the end of life" (Radbruch et al. 2020). The goal of palliative care is to improve the quality of life of patients, their families and their caregivers. The WHO (2002), following Cicely Saunders, has defined palliative care as care that includes treatment and care directed toward "total pain." Thes four types of pain include physical, psychological, psychosocial, and existential/spiritual pain. In order to address these four aspects of pain, a holistic approach to improving the patient's quality of life is needed. Furthermore, research points to the importance of spiritual and existential issues for patients at the end of life and the close relationship between their spiritual, social, and psychological well-being, which highlights the interrelation of these issues (Rego and Nunes 2019).

While patients have many needs at end of life, there is in particular a need to pay attention to their affective needs. Many patients feel strong emotions difficult to process and express when they reach this stage of the care journey, including anguish, fear, depression, anger, and dissatisfaction with their lives and in the prospect of their death (García-Navarro et al. 2021). Furthermore, research shows that persons living with advanced disease report having hidden pain and suffering not recognized by their families or healthcare providers, which leaves them feeling isolated (Quinn 2018). Spiritual care practices may at least partly help relieve this emotional distress by providing a forum for discussion and coping strategies (Rego and Nunes 2019).

Given how important spiritual care is to patients, this suggests that all healthcare providers who have contact with a patient should have some general training and awareness of it. While it is commonplace for hospital chaplains to be solicited at the end of life, spiritual care cannot be limited to only this resource person. Let us therefore look at a healthcare professional not necessarily or always seen as a spiritual guide, namely the palliative care nurse. This is a natural profession to focus on as a movement to include spiritual care in nursing has grown steadily since the 1980s, as part of advocacy for greater holistic care provision (Taylor 2003). For many nurses today, paying attention to the person's spiritual needs—whether or not the person is a believer—is an integral part of nursing practice. Miller et al. (2023) warn that it is important spiritual care does not function as simply another item on the to-do-list, but rather as an important part of the nursing methodology as a whole, as it involves listening to and cultivating a relationship with the patient, showing compassion, and fostering trust.

Furthermore, the patient's request for spiritual care may very well fall to the person in which the patient feels the most comfortable—or who simply enters the room at the right time. That person may very well be a nurse. A common situation lived in nursing practices has been detailed by Miller et al. (2023): *"At the end of a shift, you go in to check on Tracy, a 53-year-old woman with metastatic lung cancer. She is crying. She says she doesn't know how she's going to get through her next treatment, and she's terrified of dying. She looks at you and asks, 'What do you think happens after you die?'"* This example shows the right person in that moment was the person in the room, and that person happened to be the nurse, not the hospital chaplain. To leave the patient in this situation as the nurse did not feel comfortable answering these questions, or to refer them simply at a later time to the chaplain, would do little to relieve the emotional distress of the patient, particularly because the request came from the patient to this particular healthcare provider.

However, given that healthcare providers who have other roles than spiritual care may be asked to provide it, we should ask what posture they should adopt with their patient. In particular, should their spiritual care provision be *explicit* (directly addressed by the nurse or another healthcare provider to determine the spiritual needs of the healthcare provider) or *implicit* (addressing spiritual needs and beliefs only when directed by the patient)? This is an important issue, because while some patients and families appreciate the nurse directly addressing their spiritual needs, others do not necessarily see it as the nurses' role (Taylor 2003). Rego and Nunes (2019), following work by Tan (1996), advocate that an implicit approach to spiritual care works best in the context of palliative care. In this model, providers will only address spiritual matters when solicited by patients, thereby supporting the patient's choice to describe their spiritual beliefs and practices and have it included (or not) in their healthcare. In the example from Miller et al. (2023), the authors describe a situation fitting this model, as the nurse waited for the request from the patient before addressing their spiritual needs.

However, given the current lack of training and reticence by healthcare providers to ask about spiritual matters with their patients, and the uncertainty patients may face in who to ask in times of vulnerability such as end of life, it can be questioned if this approach is most suitable. In addition, in the context of affective injustice, as spiritual needs at the end of life often involve intense emotions such as fear and anguish, more implicit strategies may hinder their expression. While the implicit strategy has the benefit of promoting a person-centered perspective based upon the person's values (thereby not imposing them), the patient may not necessarily be aware that healthcare providers may serve as a spiritual and emotional resource for them. Explicit strategies that respect individual beliefs and practices such as taking a spiritual history and following the patient's lead may help overcome these barriers. Furthermore, more explicit strategies welcome the expression and recognition of strong emotions patients and their families are experiencing at the end of life. They therefore seem more suited to helping patients and their families to find the kind of affective resources, opportunities, freedoms, and recognition to help them at the end of life.

5.4.2 *Mental Health and Marginalized Groups*

I will now turn to another difficult care environment, namely mental healthcare. I will be in particular interested in exploring how spiritual care may be more responsive to some group's values and affective needs. In the chapter on patients and affective injustice (Chapter 3), I explored the difficulties some groups may face in healthcare, in particular in the ways their affects were perceived and understood by care providers. I took the example of African American women's perceptions and use of mental healthcare to show the seriousness of this problem. As I discussed, these women should be a priority group for mental health services, but they are also those who hesitate to use them. This is due to a multiplicity of factors: (1) cultural factors that discourage the expression of vulnerability; (2) bias of healthcare providers toward them and in particular the way these women's affects are treated as pathological; (3) the lack of culturally sensitive healthcare providers, including a lack of understanding of this community's life experiences and values, including their spiritual values.

In this section, I will explore how spiritual care can be more responsive to this group's affective (and other) needs, thereby improving care for them. As could be seen from the discussion in Chapter 3, there is often a mismatch between what may be valued by the patient in their care (their spirituality, their culture) and the provision of mental care as it is currently delivered. The question therefore becomes if the provision of spiritual care may provide the kinds of affective resources and opportunities needed to improve care toward this marginalized group.

First of all, let us explore the spiritual beliefs and practices in these communities to better understand how they might perceive the integration of spiritual care. To begin with, for many African Americans, spirituality plays an important part of their lives. Nearly 79% of the population identifies as Christian and a large percentage are involved in their church communities. These institutions also play an important social, cultural, civic, and political role in these communities historically (and currently) being excluded due to racism. Because of this disenfranchisement, black churches have historically offered a number of community programs for their members. This is in particular the case of the provision of mental health services, which Black churches have tended to provide in a greater amount than in churches for other communities. These churches therefore represent both a source of healing/support in line with religious beliefs and practices, but also fill the gap in inadequate care provision (Nguyen 2020). Furthermore, empirical evidence has documented the relationship between church attendance and more positive mental health for these communities, as they represent a safe space for expression, healing, and validation (Nguyen 2020). For this community, churches therefore represent an important mental health resource outside of the traditional clinical or hospital setting. This is crucial, as African Americans have been found to have increased risk of poor outcomes in mental health care due to racial disparities, including problems of treatment engagement and retention, overdiagnosis of mental illness, and overprescription of medications (Hays and Aranda 2016).

From this perspective, we can see the importance of spiritual beliefs and practices for this community and the role churches may have in fostering mental health care. However, this is not to assume that all spiritual beliefs and practices may be beneficial: if a person experiences religious doubt for example, they may feel isolated in a community that does not seem to have the same doubts. However, for those with more positive relationships, these beliefs and practices may provide the kind of resources and support they need, including for their mental health needs.

We can now understand a fourth reason these women may not choose to solicit mental health services, in particular if they feel supported by their community and their church. However, that does not mean we should necessarily "outsource" this kind of support to churches alone. First of all, churches cannot prescribe medication nor provide the kind of trained counselling led by psychologists and psychiatrists. Secondly, the poor outcomes, conflictual relationships, and poor diagnosis faced by this community shows that healthcare institutions should take a hard look at how to improve care for all communities. If churches are filling this gap, it is also because healthcare systems and structures are insufficient and not adapted to their needs. Finally, in relation to the topic of affective injustice, the problems faced by these women in having their affects be heard and understood is not only a problem in healthcare, but also among their communities, where the superwoman role has been historically ingrained. This means that churches may provide a safe space for some issues, but not necessarily for the expression of affects. There is therefore great potential for healthcare institutions to provide a safe space for those affects considered taboo elsewhere.

Therefore, if we take into consideration the values and needs of patients, it seems promising to integrate spiritual care into mental health practices to better respond to the needs of these patients and provide more culturally sensitive care. While there are many possibilities here, this might mean: (1) making links between the church and mental health services to better coordinate support and resource persons; (2) integrating community members into healthcare services to ensure care is adapted; (2) training in cultural sensitivity to help all healthcare providers provide a more welcoming space. A review article by Hays and Aranda (2016) proposes several methods with the potential to serve as affective resources, which I will detail below.

- **Group psychotherapy, with a focus on spiritual struggles and developing coping resources**

Group psychotherapy is a proposition by Bowland et al. (2012). In this method, group therapy can be a facilitator to develop positive coping strategies in line with the person's beliefs and practices. The goal of these groups is to create open dialogue and sharing of resources, rather than to prescribe answers, which lends support to the idea that these types of groups can be used across religious traditions. While a long process, the method includes discussion of the person's spiritual histories in relation to traumatic experiences; thinking about spiritual gifts, such as forgiveness or trust, as values important to the person's life and recovery; and development of a spiritual recovery plan, including with information on how spiritual coping may be good for health. The harm of negative coping, such as having a negative relationship with a higher power, is also openly discussed.

In these groups, affects take a central place. Psychological struggles with negative emotions such as anger, fear, powerlessness, and shame are openly discussed, as well as more positive emotions such as hope and how to foster it. Methods such as prayer, music, poetry, and reading give an open forum for affects to be integrated into these discussions. In this way, providing spirituality-based resources can both respond to the person's values and beliefs but also serve as a safe space for affects, including negative affects, to be openly expressed and recognized. By having a group format, it also allows others to benefit from these sharing experiences. This may be particularly important for women in need of mental health services who may be reluctant to share their affects (and thus their vulnerabilities). By giving a forum to express one's experiences and emotions, but also being able to hear and recognize those of others, this kind of method provides the affective resources and opportunities we need to work toward affective justice for this group.

- **Ensuring experts are culturally competent and promoting sharing circles**

Let us now turn to a second idea, which Crewe (2006) developed to improve mental health for older African Americans, called the *Joy of Living Program*. The goal of the programme is to improve the mental health of older persons in these communities, including increasing their abilities to talk to the doctor about their possible mental health problems. It is also geared toward improving this community's knowledge about mental health and the resources available to them. Their method uses culturally competent presenters, in particular African American experts, to tailor the program to the group's needs and integrate culturally relevant leaders. According to the authors, using African American presenters helps to address the concern of racial bias and insensitivity in questions of mental health, but also ensures faith-based linkages in the discussions.

While affects are not an explicit goal of the meetings, according to the author, such methods and the choice of culturally competent facilitators, "freed them to talk about their unique experiences free of shame or embarrassment." Their sharing circles framework enables a collective discussion, which include both listening to others, but also trying to help other members. Crewe et al. describe these circles as safe spaces for peer support, particularly as they enable participants to listen to and relate to affective experiences of other group members. It also helps these members to feel less isolated. In one example, they describe how, "one member patiently explained to someone that she had gone through the same things and felt the same way at one time. The other participants burst out in tears because she was so relieved that she was not the only one having the problem." As we can see, this resource provides a safe space for affects to be recognized and understood, when facilitated by culturally-sensitive methods and experts. In this way, they are able to be responsive to this community's affective (and other) needs, notably by actively encouraging their expression and recognition.

- **Using examples from religion to illustrate concepts and promote safe spaces**

A final example we will highlight was developed by Suite et al. (2007). It consists of adapting and implementing curriculum for healthcare professionals who provide trauma support to marginalized communities. The curriculum was developed based upon spiritual beliefs and concepts familiar to participants. They notably link the word trauma to common cultural and religious references, such as those from the Protestant Bible. When discussing affects such as fear, anxiety, and grief, they invite participants to draw theological comparisons to put them into a wider context and understanding. In this way, the curriculum was designed to help healthcare providers support those with spiritual beliefs find coherence between their lived experiences (including emotions such as fear) and their religious traditions. Their "Fear to Faith Program" is culturally sensitive as it adopts the language of trauma, including affects such as grief, in the specific traditions and canons of the religious and cultural groups of the participants. This is an example of a culturally sensitive curriculum that provides the space for individualized care adapted to the person's specific values, but also one that values individual spiritual values and practices. There is also the provision of a safe space for affects to be recognized and understood—from the community's perspective and adapted for their culture. While specifically built for those following the Protestant Bible, it will be possible to adapt such methods to other religious beliefs and practices.

As we can see from these examples, it is feasible to provide resources that pay attention to the person's community, their spiritual needs and beliefs, and in this way, to provide affective opportunities and recognition in line with their values. While literature on spiritual care highlights the challenges for healthcare providers in working with people from diverse backgrounds and cultures, we can also see from this discussion how materials may be designed and adapted to the person's individual spiritual beliefs or across religious traditions. Successful methods include in particular integrating persons from their communities as resource persons, as well as benefitting from peer resources, such as sharing circles. If designed and implemented in such ways, we will be providing both the kind of affective recognition we are after—but also start to heal the historical legacy of exclusion faced by these communities in healthcare.

5.4.3 Spiritual Care for Healthcare Providers

In this final section, I would like to take a moment to pay attention to the spiritual and affective needs of healthcare providers, which are rarely taken into account in healthcare institutions today. I will discuss examples from the COVID crisis, as in some healthcare institutions, spiritual care became an important resource for healthcare providers to deal with the moral distress they were facing. While I will take the example of the pandemic, it should be recognized that moral distress occurs (also) in non-pandemic times, suggesting a need for spiritual care in normal practice for healthcare providers. However, as the pandemic was both a difficult and creative time for those providing spiritual care, and in particular for hospital chaplains who had to deal with restrictions and at times provide remote work, it is an example to help better understand the diverse forms spiritual care can take to support healthcare providers.

To have a global view of this subject, I will first discuss the results of an International Survey of Chaplain Activity and Experience, which collected information on spiritual care provided during the pandemic in Europe, North America, Australia (and to a minimal extent in Asia, Africa, and South America) by hospital chaplains (Tata et al. 2021). Nearly 2000 persons participated in this survey, which took place during the early phases of the pandemic (May 2020). While the majority of chaplains participating in the study were Christian (Protestant or Catholic), other religious traditions including Buddhist, Muslim, and other faiths did participate to a certain extent in the survey (appropriately 9%). The survey showed that chaplains expanded their role during the pandemic, which brought with it a greater appreciation and knowledge of their work and of the kind of support they could provide to healthcare professionals. Chaplains highlighted in particular an increased need among staff to talk to the hospital chaplain about personal experiences, concerns and affects, including fear for the future, for their health and those of others. As we can see from this study, affects were at the heart of the healthcare provider's experiences during this time and they often turned to the hospital chaplain for help.

The affective support provided by the chaplain during the pandemic has also been highlighted in another study, which showed how they provided a safe space for healthcare providers to openly discuss their fears, loneliness, and stress. This qualitative study, led by Domaradzki (2022) in Poland, quoted one chaplain's work as saying: *"spiritual care also covers the personnel...I support them, bring them hope during isolation from their families and assist in their problems and fears."* The study reported that healthcare staff (including doctors and nurses) approached the chaplain to ask for their help in managing their emotions resulting from increased workload, worries about their loved ones, or mourning after patients. According to the author, chaplains emphasized that professionals needed to talk about difficult emotions and the chaplain could be a key resource person. For instance, according to one person interviewed, *"my contact with the clinicians is always very moving. Physicians and nurses also need compassion and conversation. Even asking a simple question 'How do you feel?' or 'It's good you are here' means something, especially for overloaded nurses. They really appreciate my presence."* Another said, *"one surgeon came to me and asked for a talk. He said he had problems with expressing his emotions. I remember his words. He said: 'I feel overwhelmed by what is going on, by the number of the sick and by the fact that we can save only a few patients.'"* As we can see from this testimony, chaplains functioned as an affective resource in which healthcare providers—whether they be nurses, doctors, or other healthcare professionals—could turn to in order to talk through and process their emotions.

During this time, chaplains also innovated and found ways to support staff. This included creating wellness surveys, debriefs with staff in the service, providing educational resources on self-care, anxiety and stress, as well as more informal moments of exchange such as providing "wellbeing hubs" in the hospital or blessing jars in which staff could express words of encouragement and reflection (Tata et al. 2021). As we can see from these examples, hospital chaplains provided affective opportunities (methods encouraging affective expression) for staff through both formal and informal exchanges.

While this kind of research shows the important role chaplains played during the pandemic and their role in helping to relieve the moral distress of professionals, this does not mean the pandemic was necessarily an ideal period for chaplains to provide spiritual care. Some chaplains were asked to work remotely, or to avoid contact with healthcare professionals, families, or patients, during the pandemic, which impeded their abilities to provide what they conceived of as good spiritual care (Snowden 2021). Their definition of "good" spiritual care notably included the unplanned interactions which allowed them to listen to the healthcare provider's "joys and complaints" and promote their wellbeing in informal ways, such as informal discussions in the hallway about their day. However, despite these restrictions, they also showed creativity in adapting their services, such as providing a "hotline" for staff to call them or making a newsletter to support staff. Despite these creativities, the request to work remotely shows their marginal place at the hospital, at a time in which only "essential workers" were needed on site. The lack of recognition placed a considerable strain on hospital chaplains during this time, as it questioned their place in the care plan.

In addition, while hospital chaplains were involved in spiritual and emotional care to patients, families, and healthcare providers during this time, they had little affective resources and opportunities for themselves. As the study by Domaradzki (2022) shows, the pandemic was also an emotionally challenging time for them and they experienced intense affects including sadness, anxiety, fear, and frustration. These feelings were in particular related to the moral distress they felt in being excluded in care for dying patients, who they felt were suffering and dying alone due to visitor restrictions. They were also distressed about their inabilities to provide the kind of spiritual care they understood as best fitted to the needs of patients, families, and healthcare providers, which meant both spontaneous and planned spiritual care that could only take place in-person. This finding suggests that while spiritual care became an important means for healthcare actors to survive during the pandemic, we did little to support chaplains and other support persons who provided other types of essential care during this time.

To conclude this section, the pandemic gives a number of lessons in providing affective resources and opportunities for healthcare providers through spiritual care: (1) providing safe spaces for affects needs dialogue and confidence-building, which is difficult (but not impossible) to do remotely, in particular for healthcare providers who may be reluctant to express their affects; (2) diverse methods can be used to create affective safe spaces, including blessing jars, planned groups, informal moments of exchange in hallways; (3) in order to permit affective resources and recognition, a dedicated person (such as a chaplain) who is considered a "neutral" resource may be helpful; (4) we will also need to take care of the affective needs of hospital chaplains to enable them to provide quality care for patients, their families, and healthcare providers.

5.5 The Delegation of Affects to Spiritual Care Providers: A Necessary Caution

Before concluding this chapter, it will be necessary to nuance the debate on the possibility of giving a greater place for affects through spiritual care. As I have shown, the inclusion of the hospital chaplain brings several ways to work toward affective justice, notably through the provision of a "non-healthcare provider" into the hospital space, permitting other ways of caring to be possible, but also by the methods of spiritual care that actively seek to bring out and work with affects. However, because the hospital chaplain is often seen as the person with the time, resources, listening, compassion, and training to deal with the patient's affects, there is also a tendency to "delegate" the emotional work to this person. According to Etienne Rochat, a hospital chaplain at Lausanne University Hospital, in the case of somatic illness, those emotions which are strong/disruptive are delegated to the psychologist or psychiatrist; whereas those affects which are vague, less intense, or less strong will be delegated to the hospital chaplain, who is seen as the person with the time and the listening skills to handle them, at least in the hospital space, where there is regularly a chaplain present. As I have been advocating throughout this chapter, a diversity of healthcare professionals need to be trained in spiritual care practices in order to provide sufficient affective resources and opportunities for the diverse needs of patients and their families. This means providing affective resources and opportunities cannot be limited to the hospital chaplain.

5.6 Conclusion

In this chapter, I have explored the interlinkage of spiritual care and affects. On the one hand, spiritual beliefs and practices may function as an individual affective resource and opportunity for patients when dealing with an illness, notably as it may enable persons to have greater well-being and to make sense of their illness. On the other hand, this does not mean this has to be (only) in the private domain. In this chapter, I have explored the potential of spiritual care as a means to provide the affective resources, opportunities, freedoms, and recognition in ways that are rarely possible with the other methods used in the hospital. I have also shown for those with spiritual beliefs and practices, their inclusion helps them feel supported in their illness journey and to accept treatment plans in line with their values. However, even for those non-believers, it may provide affective resources and opportunities that can change how healthcare is practiced and the place for affects within these spaces. From blessing jars, to discussion groups with a chaplain, to sharing circles among patients, spiritual care can both help those with and without beliefs in expressing and being recognized. As seen in the previous chapter on narrative methods, while affects are not necessarily the explicit goal, they naturally come into dialogue with those willing to encompass a wider view of care. Spiritual care can provide that needed sidestep to allow other ideas and ways of being and caring to the come to the forefront, including a greater place for affects.

As I have been advocating throughout this chapter, it will be important not to impose beliefs or practices on healthcare actors. It will also be vital to individualize spiritual care in function of the person's relationship to their spiritual beliefs and their communities. While this gives a large task to those practicing spiritual care, it is the means by which we can work toward the ideal of person-centered care, one that is equitable and responsive to the individual person's values and beliefs. From a practical standpoint, it also does not demand significant resources. The means to guarantee spiritual care in line with the person's values often involves simple methods—such as taking a spiritual history, or recognizing spiritual beliefs and practices in group discussions. These strategies may not add much time to the consultation, but it can make a world of difference to the patient.

References

Astrow, Alan B., Ann Wexler, Kenneth Texeira, M. Kai He, and Daniel P. Sulmasy. 2007. "Is Failure to Meet Spiritual Needs Associated With Cancer Patients' Perceptions of Quality of Care and Their Satisfaction With Care?" *Journal of Clinical Oncology* 25 (36): 5753–5757. https://doi.org/10.1200/JCO.2007.12.4362

Augustyn, Brian D., Todd W. Hall, David C. Wang, and Peter C. Hill. 2017. "Relational Spirituality: An Attachment-Based Model of Spiritual Development and Psychological Well-Being." *Psychology of Religion and Spirituality* 9 (2): 197–208. https://doi.org/10.1037/rel0000100

Beck, Richard. 2006. "God as a Secure Base: Attachment to God and Theological Exploration." *Journal of Psychology and Theology* 34 (2): 125–132. https://doi.org/10.1177/009164710603400202

Borneman, Tami, Betty Ferrell, and Christina M. Puchalski. 2010. "Evaluation of the FICA Tool for Spiritual Assessment." *Journal of Pain and Symptom Management* 40 (2): 163–173. https://doi.org/10.1016/j.jpainsymman.2009.12.019

Bowland, S., T. Edmond, and R. D. Fallot. 2012. "Evaluation of a Spiritually Focused Intervention with Older Trauma Survivors." *Social Work* 57 (1): 73–82. https://doi.org/10.1093/sw/swr001

Crewe, Sandra Edmonds. 2006. "Joy of Living: A Community-Based Mental Health Promotion Program for African American Elders." *Journal of Gerontological Social Work* 48 (3–4): 421–438. https://doi.org/10.1300/J083v48n03_09.

Domaradzki, Jan. 2022. "'We Are Also Here'—Spiritual Care Practitioners' Experiences of the COVID-19 Pandemic: A Qualitative Study from Poland." *Journal of Religion and Health* 61 (2): 962–992. https://doi.org/10.1007/s10943-021-01492-3

García-Navarro, E. Begoña, Alicia Medina-Ortega, and Sonia García Navarro. 2021. "Spirituality in Patients at the End of Life—Is It Necessary? A Qualitative Approach to the Protagonists." *International Journal of Environmental Research and Public Health* 19 (1): 227. https://doi.org/10.3390/ijerph19010227

Gallegos, Francisco. 2021. "Affective Injustice and Fundamental Affective Goods." *Journal of Social Philosophy* n/a (n/a). https://doi.org/10.1111/josp.12428.

Gijsberts, Marie-José H. E., Anke I. Liefbroer, René Otten, and Erik Olsman. 2019. "Spiritual Care in Palliative Care: A Systematic Review of the Recent European Literature." *Medical Sciences* 7 (2): 25. https://doi.org/10.3390/medsci7020025

Hall, Eric J., and Richard A. Powell. 2021. "Valuing the Spiritual." *Journal of Religion and Health* 60 (3): 1430–1435. https://doi.org/10.1007/s10943-021-01206-9.

Hall, Todd W. 2004. "Christian Spirituality and Mental Health: A Relational Spirituality Paradigm for Empirical Research." *Journal of Psychology & Christianity* 23 (1): 66–81.

Hays, Krystal, and Maria P. Aranda. 2016. "Faith-Based Mental Health Interventions With African Americans: A Review." *Research on Social Work Practice* 26 (7): 777–789. https://doi.org/10.1177/1049731515569356

Hvidt, Niels Christian, Kristina Tomra Nielsen, Alex K Kørup, Christina Prinds, Dorte Gilså Hansen, Dorte Toudal Viftrup, Elisabeth Assing Hvidt, et al. 2020. "What Is Spiritual Care? Professional Perspectives on the Concept of Spiritual Care Identified through Group Concept Mapping." *BMJ Open* 10 (12): e042142. https://doi.org/10.1136/bmjopen-2020-042142

Kim-Prieto, Chu, ed. 2014. *Religion and Spirituality Across Cultures*. Vol. 9. Cross-Cultural Advancements in Positive Psychology. Dordrecht: Springer Netherlands. https://doi.org/10.1007/978-94-017-8950-9

Kirchoff, Robert W., Beba Tata, Jack McHugh, Thomas Kingsley, M. Caroline Burton, Dennis Manning, Maria Lapid, and Rahul Chaudhary. 2021. "Spiritual Care of Inpatients Focusing on Outcomes and the Role of Chaplaincy Services: A Systematic Review." *Journal of Religion and Health* 60 (2): 1406–1422. https://doi.org/10.1007/s10943-021-01191-z

Klitzman, Robert, Jay Al-Hashimi, Gabrielle Di Sapia Natarelli, Elizaveta Garbuzova, and Stephanie Sinnappan. 2022. "How Hospital Chaplains Develop and Use Rituals to Address Medical Staff Distress." *SSM—Qualitative Research in Health* 2 (December): 100087. https://doi.org/10.1016/j.ssmqr.2022.100087

Koenig, Harold G. 2012. Religion, Spirituality, and Health: The Research and Clinical Implications. *ISRN Psychiatry* 2012: 1–33. https://doi.org/10.5402/2012/278730

Koenig, Harold G., Kathleen Perno, and Ted Hamilton. 2017. "The Spiritual History in Outpatient Practice: Attitudes and Practices of Health Professionals in the Adventist Health System." *BMC Medical Education* 17 (1): 102. https://doi.org/10.1186/s12909-017-0938-8

Lalani, Nasreen, and Amity Chen. 2021. "Spirituality in Nursing and Health: A Historical Context, Challenges, and Way Forward." *Holistic Nursing Practice* 35 (4): 206–210. https://doi.org/10.1097/HNP.0000000000000454

Larson, David B., and Susan S. Larson. 2003. "Spirituality's Potential Relevance to Physical and Emotional Health: A Brief Review of Quantitative Research." *Journal of Psychology and Theology* 31 (1): 37–51. https://doi.org/10.1177/009164710303100104

Lucchetti, Giancarlo, Rodrigo M. Bassi, and Alessandra L. Granero Lucchetti. 2013. "Taking Spiritual History in Clinical Practice: A Systematic Review of Instruments." *EXPLORE* 9 (3): 159–170. https://doi.org/10.1016/j.explore.2013.02.004

MacLean, Charles D., Beth Susi, Nancy Phifer, Linda Schultz, Deborah Bynum, Mark Franco, Andria Klioze, Michael Monroe, Joanne Garrett, and Sam Cykert. 2003. "Patient Preference for Physician Discussion and Practice of Spirituality: Results from a Multicenter Patient Survey." *Journal of General Internal Medicine* 18 (1): 38–43. https://doi.org/10.1046/j.1525-1497.2003.20403.x

Miller, Megan, Katie Addicott, and William E. Rosa. 2023. "Spiritual Care as a Core Component of Palliative Nursing." *AJN, American Journal of Nursing* 123 (2): 54–59. https://doi.org/10.1097/01.NAJ.0000919748.95749.e5

Monod, Stefanie M, Etienne Rochat, Christophe J Büla, Guy Jobin, Estelle Martin, and Brenda Spencer. 2010. "The Spiritual Distress Assessment Tool: An Instrument to Assess Spiritual Distress in Hospitalised Elderly Persons." *BMC Geriatrics* 10 (1): 88. https://doi.org/10.1186/1471-2318-10-88

Nguyen, Ann W. 2020. "Religion and Mental Health in Racial and Ethnic Minority Populations: A Review of the Literature." Edited by Laura P Sands. *Innovation in Aging* 4 (5): igaa035. https://doi.org/10.1093/geroni/igaa035.

Puchalski, Christina, and Anna L. Romer. 2000. "Taking a Spiritual History Allows Clinicians to Understand Patients More Fully." *Journal of Palliative Medicine* 3 (1): 129–137. https://doi.org/10.1089/jpm.2000.3.129

Quinn, Barry. 2018. "Making Sense of Pain and Loss: Searching for Meaning While Living with Cancer." *Cancer Nursing Practice* 17 (5): 29–36. https://doi.org/10.7748/cnp.2018.e1521

Radbruch, Lukas, Liliana De Lima, Felicia Knaul, Roberto Wenk, Zipporah Ali, Sushma Bhatnaghar, Charmaine Blanchard, et al. 2020. "Redefining Palliative Care—A New Consensus-Based Definition." *Journal of Pain and Symptom Management* 60 (4): 754–764. https://doi.org/10.1016/j.jpainsymman.2020.04.027

Rego, Francisca, and Rui Nunes. 2019. "The Interface between Psychology and Spirituality in Palliative Care." *Journal of Health Psychology* 24 (3): 279–287. https://doi.org/10.1177/1359105316664138

Selman, Lucy Ellen, Lisa Jane Brighton, Shane Sinclair, Ikali Karvinen, Richard Egan, Peter Speck, Richard A Powell, et al. 2018. "Patients' and Caregivers' Needs, Experiences, Preferences and Research Priorities in Spiritual Care: A Focus Group Study across Nine Countries." *Palliative Medicine* 32 (1): 216–230. https://doi.org/10.1177/0269216317734954.

Snowden, Austyn. 2021. "What Did Chaplains Do During the Covid Pandemic? An International Survey." *Journal of Pastoral Care & Counseling: Advancing Theory and Professional Practice through Scholarly and Reflective Publications* 75 (1_suppl): 6–16. https://doi.org/10.1177/1542305021992039

Suite, Derek H., Stephen A. Rollin, J. C. Bowman, and Robert D. La Bril. 2007. "From Fear to Faith: Efficacy of Trauma Assessment Training for New York–Based Southern Baptist Church Groups." *Research on Social Work Practice* 17 (2): 258–263. https://doi.org/10.1177/1049731506296678

Tan, Siang-Yang. 1996. "Religion in Clinical Practice: Implicit and Explicit Integration." In *Religion and the Clinical Practice of Psychology.*, edited by Edward P. Shafranske, 365–387. Washington: American Psychological Association. https://doi.org/10.1037/10199-013

Tata, Beba, Daniel Nuzum, Karen Murphy, Leila Karimi, and Wendy Cadge. 2021. Staff-Care by Chaplains during COVID-19. *Journal of Pastoral Care & Counseling: Advancing theory and professional practice through scholarly and reflective publications* 75: 24–29. https://doi.org/10.1177/1542305020988844

Taylor, Elizabeth Johnston. 2003. "Nurses Caring for the Spirit: Patients With Cancer and Family Caregiver Expectations." *Oncology Nursing Forum* 30 (4): 585–590. https://doi.org/10.1188/03.ONF.585-590

Van Cappellen, Patty, Ruixi Zhang, and Barbara L. Fredrickson. 2023. "The Scientific Study of Positive Emotions and Religion/Spirituality." In *Handbook of Positive Psychology, Religion, and Spirituality*, edited by Edward B. Davis, Everett L. Worthington, and Sarah A. Schnitker, 315–328. Cham: Springer International Publishing. https://doi.org/10.1007/978-3-031-10274-5_20

Vishkin, Allon, Pazit Ben-Nun Bloom, Shalom H. Schwartz, Nevin Solak, and Maya Tamir. 2019. "Religiosity and Emotion Regulation." *Journal of Cross-Cultural Psychology* 50 (9): 1050–1074. https://doi.org/10.1177/0022022119880341

World Health Organization. 2002. *National cancer control programmes: policies and managerial guidelines.* World Health Organization.

Chapter 6
Affective Justice, Health Democracy, and Citizen Engagement

Abstract This chapter will consider to what extent patient engagement can move us toward affective justice. I will in particular explore the potential of increased patient participation to provide affective resources, opportunities, and recognition for patients. This may happen in the consultation, via the integration of patient partners/advocates in healthcare delivery, and with the implication of patient associations. While the chapter is focused mainly on the potential of this methodology for patients, I will also consider how the greater involvement of citizens may provide some new safe spaces for healthcare providers. While I will argue for the potential of these new forms of health democracy as a way to work toward affective justice, I will also pay close attention to the risks for patients and healthcare providers in their inclusion, notably the potential risk of experiencing affective injustice because of their involvement. In the last section of this chapter, I will also focus on the affective injustice that patient partners/advocates may experience in these new forms of engagement.

Keywords Affects · Patient advocate · Patient partner · Patient expert · Health democracy · Patient engagement

6.1 Introduction

Having discussed two promising methods in the previous chapter (narration and spirituality) that can help work toward affective justice, I will now turn to a discussion on how fellow citizens can help move the discussion forward. Following Gallegos (2021), I am interested in is finding affective resources, opportunities, freedoms, and recognition for patients, their families, and healthcare providers. I will explore how various forms of citizen engagement may help to achieve these goals. As I will show, there are promising reasons to believe that fellow citizens can serve as resource persons for affective sharing, as well as provide new opportunities for affects to be recognized in healthcare teams and institutions. My goal in this chapter is to provide a nuanced perspective, one in which we recognize the possibilities for fellow citizens to honor, hear, and recognize patient affects, but also show the limits (and potential

B. Bogaert, *Affective Injustice in Healthcare*, The International Library of Bioethics 112, https://doi.org/10.1007/978-3-031-94378-2_6

dangers) of having them speak for others—and decide which affects should be heard and recognized.

Before beginning the discussion, an introduction is in order of what I mean by the terms patient engagement, health democracy, and patient partner/advocate. First of all, the term health democracy refers to the active involvement of citizens in the development and implementation of healthcare policy, in a spirit of dialogue and consultation (Arveiller and Tizon 2016). It is a global term to encompass the different ways citizens participate in healthcare policy, advocacy, and research. Today, there are multiple ways in which citizens participate in health democracy, going from the individual patient-provider relationship to participating in policy making and service delivery. These include: (1) the engagement of the individual patient in the singular colloquy of the patient-healthcare provider relationship, most notably in shared decision-making; (2) those working directly in the hospital as patient partners/advocates; (3) citizens participating at the strategic level as patient representatives in policy or ethical committees; (4) those who work as patient research partners in research projects; (5) the peer support and advocacy provided through patient associations; (6) citizens working as patient teachers, in particular in medical schools. While shared decision making in the consultation has been advocated since at least the 1960s, these new tasks for patients in hospital organizations and research have emerged since the 1980s.

For the purposes of clarity, in this chapter, I will refer to these new roles (beyond that of shared decision-making in the consultation) as roles of patient partnership or advocacy. These are the two terms widely accepted in the English-speaking context, although patient expert or patient advisor (or other terms) are also used depending on their specific function and role. In this chapter, I will in particular focus on the first (shared decision-making), second (hospital engagement) and fifth (patient associations) forms of engagement, as they are the most relevant to the discussion on achieving affective injustice for patients in their healthcare. However, it is to be acknowledged that citizens bring significant epistemic contributions to these other forums, which has been widely documented in literature (Skovlund et al. 2020; Bombard et al. 2018; Ali et al. 2006; Boote et al. 2010 etc.). In Chapter 8, I will also discuss their contribution to medical education.

Having clarified these terms, let us now explore the term citizen engagement and what it may represent. I will use this term to refer to how health democracy happens in practice (namely through citizen engagement in the various forums I have stated above) but also how the patient participates at the clinical level in decisions about their healthcare. Higgins et al. (2017, p. 31) propose that patient engagement be understood as, "*the desire and capability to actively choose to participate in care in a way uniquely appropriate to the individual in cooperation with the healthcare provider or institution in the interests of maximizing outcomes or experiences of care.*" As we can see from this definition, we cannot "engage" alone, we need someone to engage *with.* It is a question of promoting a partnership: this may take place in the consultation (between the healthcare provider and the patient) but it may also encompass the new forms of health democracy detailed above, in which a patient partner/advocate may engage with a healthcare institution outside of their healthcare in their service toward other patients. My choice of the term "engagement" also reflects the realities of health democracy, in that citizens are free to "disengage" from these forms of participation should they wish (Pierron 2006). Given their uncertain place in healthcare institutions and largely voluntary roles, many citizens feel free to disengage, particularly when they themselves are subject to affective and epistemic injustice, as I will discuss.

As we can see, health democracy and citizen engagement are based up on an ethical ideal. They envisage patients as persons with knowledge to be shared with others, and an expectation of recognition of this knowledge by healthcare providers, the public, and institutions. Patient engagement is an important subject in the epistemic justice discussion. Reaching this ideal means having what is often seen as a passive actor (the patient) being heard, recognized, and understood in knowledge gathering related to healthcare policy and implementation, including by contributing to new collective epistemic resources.

Beyond the ethical ideal, however, it should be recognized these new roles are defended and developed for other reasons, in particular a number of practical issues from the public health perspective. First of all, there is the belief that by integrating the experiential knowledge of patients into healthcare planning, implementation, and policymaking, care will be improved. Their contribution is notably thought to minimize unnecessary costs that may be caused by treatment inadherence or conflictual doctor-patient relationships. Furthermore, in a democratic ideal, their involvement has also been advocated as way to increase public trust and engagement. While more empirical research is needed to clarify the patient partner/advocate's epistemic contribution (Bogaert 2023) and in what means they improve care delivery (Bombard et al. 2018), these are promising reasons to believe that the integration of citizens can help improve the healthcare journey and quality of care. However, the question that will be most relevant to our discussion is how citizen participation facilitates or hinders affective justice for other patients or healthcare providers. As we will see, these questions are often interlinked, as by facilitating safe affective spaces, we will also likely improve care relationships and quality of care.

Before moving forward with the discussion, it will be necessary to take a moment to understand to what extent health democracy has actually become a reality, as there is often a gap between an ideal and its implementation. At the time of writing, the level of development and integration of citizens in health democracy varies widely across countries (Vanstone et al. 2023). It is to be noted in particular that the greatest development has been documented in English-speaking countries (Canada, United States, Australia, United Kingdom) and in Europe (Germany in particular, but increasingly in other European countries). It has been slower to emerge elsewhere, although it is rapidly becoming a priority. While this might give the impression that health democracy is solidly in place (at least in these countries), it is to be acknowledged that even where health democracy is well developed, citizens continue to face resistances to their inclusion. This resistance is both individual (for instance a doctor who feels ambivalent about a patient partner participating in a coordination meeting) but also institutional, as their expertise is rarely formalized in the form of a work contract.

This resistance is linked to the uncertainty of their place and what they can bring to improve patient care, but also ongoing resistances to their inclusion due to hierarchies. Take for example the controversy over what term can describe their role. Should it be as a patient advocate, patient expert, patient partner, or some other term? There continues to be resistance to calling a citizen an "expert" as it puts them on par with the "expert" doctor. While there is increasing advocacy to recognize the expertise of the patient, gained from their personal experience with the illness and their treatment, it is different in form and content to the expertise acquired through medical training. Thus, both patient partners/advocates and healthcare professionals have moved away from some terms which cause tension in order to focus on more neutral terms. At the current time, terms such as patient partner or patient advocate have been increasingly used: however, these choices also show that patient engagement remains far from achieved.

Furthermore, while citizen engagement in some forums—in particular in research, public policy, and teaching—is rapidly developing, other spaces of citizen engagement face more resistance than others. For instance, in France, where this researcher has lived and worked for many years, patient teachers (*patients formateurs/enseignants*) are rapidly being integrated into medical teaching, most often in classes alongside a doctor or social sciences lecturer. They are also starting to be integrated into health research projects, although they continue to face significant resistance from researchers. However, their inclusion in hospital teams as a patient partner/advocate continues to be fraught with difficulty. At the time of writing, there remain few (if any) examples of successful collaboration of this type in France. They most often serve in a voluntary capacity and often face considerable resistance due to hierarchies. Their work is also strictly limited so as not to "interfere" with healthcare delivery or to put into difficulty other healthcare professionals' expertise. France is not the exception, as citizens face similar kinds of resistances in other countries as well. The uncertain place for patient partners/advocates in the hospital was never more evident than during the COVID crisis. During this time, around the world, patient partners/advocates were excluded for sanitary reasons (Crogan 2021), not unlike the hospital chaplains discussed in the last chapter. This should make us pause and question what is essential care for patients. As I will show in the context of affective injustice, these resource persons are an important and practical means of helping the patient's affects be better seen and recognized in our healthcare institutions. It is for this reason I advocate for their sustainable inclusion and place.

Having introduced the main terms in the debate, let us now turn to the various formulations of health democracy on the ground and in what means, spaces, and forms patient engagement may contribute to affective justice.

6.2 Health Democracy on the Ground and Affective Injustice

In this section, I will discuss three levels of citizen engagement to better understand how they may influence the understanding and recognition of affects. To begin, it will be important to analyze what may happen when patients become actively engaged in shared-decision making in the consultation, and if their affects may be better understood by the healthcare provider because of it. I will then explore the integration of patient partners/advocates as part of the healthcare team in the hospital. I will notably explore how these persons may function as affective justice resources and opportunities for patients and healthcare providers. Thirdly, I will explore the role of citizens outside of the hospital, namely in patient associations, and how they might facilitate affective justice for patients, particularly in situations where they may not feel comfortable sharing their affects in other forums. While my discussion will principally be centered on patient experiences, I will also pay attention to healthcare provider's needs and show how their integration may also encourage their affective expression.

6.2.1 Patient Engagement at the Clinical Level: The Singular Colloquy

In this section, I will get the conversation going on patient engagement by looking at what happens at the clinical level, in particular in shared decision making. I will explore how the active participation of the patient—and in return, a commitment by the healthcare provider to listen to and integrate their experiential knowledge—may help to work toward greater affective justice for patients.

Shared decision-making (SDM) is a concept that emerged in the early 1980s, although the foundations are much older, based upon ethical questionings on medical decision-making and patient-provider relationships (Dogba et al. 2016). While the term has taken on various formulations and definitions, SDM essentially involves a knowledge-giving and receiving process between a patient and a healthcare provider. In the model, it is advocated that the patient has experiential knowledge[1] about their illness, while the healthcare provider has medical knowledge and experience. These two forms of knowledge are supposed to lead to a mutual decision, in which the patient is empowered to make an informed decision about their healthcare (Elwyn et al. 2012). As we can see, shared decision making is both an ethical and epistemic issue, as it considers the ways patients will participate in healthcare decision-making by giving their testimony, with an expectation they will be treated fairly by their healthcare provider by having it heard, understood, and integrated into decision-making.

[1] By experiential knowledge, I refer to Borkman's definition (1976), which refers to truth based on personal experience with a phenomenon.

While the ideal form of shared decision-making involves an empowered patient who is able to actively participate and make decisions about their healthcare, on the ground it takes a variety of forms, depending on the individual clinician, the family, the patient and their singular relationship. It may include an active role for an "expert patient" who knows their illness very well, who is capable of exercising a clear decision based upon their experiential and scientific knowledge, including their awareness of and knowledge of medication and treatment options. However, it may also involve a less active involvement of the patient, who may consent to a procedure or treatment but in which the healthcare provider plays a greater role. In the second type of situation, patients (voluntarily or not) often delegate the decision-making responsibility to their healthcare provider. In reality, most shared decision-making is closer to the second case due to the epistemic authority of the doctor which is difficult to challenge. However, it is also be to recognised that a greater decision-making authority may also be the role expected by the patient, as they go to the healthcare provider to seek their advice and expertise about their illness.

While the first case is growing, with more calls for an active patient role, the second, more passive form remains the reality of most so-called shared-decision making today. This may be due to the continuing epistemic authority of the doctor (which may be criticized as paternalism), but we should also consider the implications of asking the patient to take an active role in decision-making. Given the considerable strain of living with an illness and its treatment, the choice to "delegate" the decision-making authority to a healthcare provider may also be a means to relieve the illness burden. From this perspective, the patient's choice of how and what ways they wish to be empowered in shared decision-making needs to be prioritized.

However, despite a lesser or greater degree of patient implication, shared decision-making involves at the very least taking seriously the knowledge the patient chooses to give to the healthcare provider and a conscious effort by the healthcare provider to hear and understand it. If we accept affects as a form of knowledge, then they are part of the experiential knowledge that the patient gives to the healthcare provider. Affects are therefore also to be taken seriously and integrated into the final decisions taken. The shared decision-making model in this light provides the means to work toward affective justice for patients. First of all, by welcoming those affects the patient chooses to express, and by having them be heard and understood, the consultation can serve as an affective safe space for the patient. Secondly, expressing affects in shared decision-making also has an epistemic value, as it may provide important information about the person's well-being and their life situation that can be taken into account in the healthcare plan. In this way, being able to openly express one's affects in the consultation gives the opportunity for patients to benefit from personalized care, whilst also contributing epistemically to the consultation.

However, despite this ideal, researchers are increasingly documenting the epistemic injustice (and we would like to add affective injustice) that happens in shared decision-making (Thomas et al. 2020). In practice, the patient's knowledge—including affects—is often not "shared," but rather ignored, misunderstood, or silenced by the healthcare provider. As understood from the overall epistemic injustice discussion, this can lead to self-silencing behaviours on the part of the patient, who knows they will not be heard or understood and who chooses to stay silent. It is also to be noted this can lead to ripple effects on the patient beyond the initial situation of epistemic and affective injustice, as well as lead to emotional disorientation that may further confuse and isolate them in their healthcare journey.

To better understand this phenomenon, let us take an example of changing or adjusting medications for older adults. This is a key moment for shared decision-making, as these patients both face complexities in the care plan due to multiple and complex medications, and also because changing or adjusting medications involves a trade-off that may be detrimental for the quality of life of the patient. Despite the importance of shared decision-making for older adults, Manias et al. (2024)'s empirical research shows that it rarely happens in practice. Decisions about medication changes are characterised by an asymmetry of patient involvement, including a lack of input about the patient's preferences, values, expectations, and beliefs. Older patients in their study were often not even given information about changes to their medications until after the decisions were made. In return, healthcare providers interviewed explicitly upheld their epistemic authority about medication decisions. While the expertise of the healthcare provider was not necessarily questioned by the authors, they demonstrated the effects on patient behaviour. They show that it notably resulted in patient resistance to taking medications, as well as insufficient disclosure of important information to healthcare professionals. The reason for this was unsurprising. Patients stated that because they were rarely involved in shared decision-making, they did not feel comfortable in consulting with healthcare providers about medication changes. This led them to make medication changes without informing their doctor. Other reasons for the lack of disclosure were that they were worried or fearful their doctor may not agree with their decision, or because they believed that their doctor would not have made the time to listen to them. We can see here the effects of epistemic injustice in creating conflictual doctor-patient relationships, as well as creating suboptimal care.

The study also showed that as healthcare providers considered themselves the epistemic authorities for this type of decision making, it was up to the patient or family member to initiate conversations about medication changes. As we can see from these research results, by excluding patients from shared decision-making, the doctor provided an unsafe epistemic and affective space, one in which patients self-silenced themselves in the consultation and took autonomous decisions outside of it. The problem is that these decisions may have serious consequences for the patient, notably because it led them to change medications without discussing it with their doctors.

What this first discussion has shown is that while shared decision-making could provide the kind of safe epistemic and affective space we are searching for, in reality the continuing hierarchy of medical expertise, conflictual patient-healthcare provider relationships, and situations of silencing and self-silencing means there is still a long way to go to achieve it. There is still much work to do at the institutional level to enable and encourage practioners to take the time to work with patients to make a safe affective and epistemic space. Let us now explore other possible methods to facilitate affective justice for patients, and in particular the possibilities offered by new forms of citizen engagement.

6.2.2 *Patient Partners/advocates at the Hospital*

In this section, I will discuss an evolving issue, namely the integration of patient partners/advocates in hospital teams. While this practice varies considerably between countries, the main idea is that a fellow patient or patient representative[2] can help other patients in the hospital, including by helping them to navigate the healthcare system, understand and accept their illness, negotiate with the healthcare provider, or just find someone they can talk to more freely. These persons also increasingly participate directly in healthcare delivery, such as serving as a fellow teacher in therapeutic patient education alongside a nurse.

While the inclusion of citizens into hospitals have been discussed and proposed for several decades, ongoing resistance to their inclusion continues. This is for several reasons. First of all, they are often seen as "interfering" in the work of healthcare professionals or as some attempt to replace them. There is also the issue of professionalization, notably to what extent these persons should be trained to take up these roles in order to achieve "objectivity" in their relationships with patients. Finally, an ongoing question is whether or not they should be considered employees (with a corresponding salary) or volunteers. The question of renumeration is a contentious one in many countries, in particular for healthcare staff such as nurses, in which the hourly salary rate proposed for patient partners/advocates is not that different from their own.

Because of these issues, at the time of writing, this type of citizen engagement continues to be among the most fragile of health democracy. There continues to be much confusion about their roles, the limits and contours of their actions, what are their responsibilities, and how they their work should be evaluated. Should they collaborate with the healthcare team in implementation, for instance by participating in therapeutic patient education? Should they instead serve an advocacy role, for instance by bringing requests or complaints to the attention of the healthcare team? Should they provide safe spaces for listening, such as helping the patient accept and understand the illness, or to navigate the opaque hospital space? Should they participate in interdisciplinary meetings with healthcare providers? There is a multiplicity of perspectives, diverse roles and responsibilities that citizens may serve (and often do) in a hospital space, whether in an official or unofficial role.

In this section, I will explore the possibilities for these persons to serve as affective resources for patients. I will first of all detail whether these persons can provide

[2] The term patient representative is often used in conjuncture with those who take leadership roles in patient associations, which I will discuss more in the next section. These persons may be patients or former patients, but they are also family members, or those connected in some way with the illness in question (such as a social worker or healthcare worker who has taken a leadership role in a patient association).

the safe space for patient's affects to be heard and understood, including in what temporalities and spaces. I will then explore whether their integration may help the patient's affects to be better integrated into decisions taken about care. Finally, I will also discuss in what situations patient partners/advocates may also be the cause of affective injustice for other patients.

In terms of the first issue, which concerns how a fellow patient may help the patient's affects to be better heard and understood, there are promising reasons to think that these persons may serve as affective resources for fellow patients. As non-healthcare staff and often former patients, they necessarily incarnate other roles and means of caring for patients and understanding care. While the methods used are diverse, they may involve making themselves available to patients to discuss their problems/issues/questions to help them better understand and navigate their healthcare journey. In this capacity, they may provide a safe space, as some patients may consider them less threatening and find them easier to talk to than their healthcare provider, particularly when they have faced similar issues, confusion, and emotions. As seen in the previous section in the example of failed shared-decision making in elderly populations, there is a recurring perception by some patients that their experiences will not be taken seriously or understood by the healthcare provider. Patients do not necessarily have this view of other patients, however, as they do not have the epistemic authority of doctors. In this way, they may provide a safe space for the patient to openly express what they are most fearful of, worried, or concerned about in ways that may not be possible for the patient in front of their doctor.

While it will be up to the patient partner/advocate to acknowledge, hear, and understand these affects, it is likely that persons may be more open to the affects the patient is experiencing—fear, anger, anxiety, hope—in particular if they have lived through similar experiences in the past. By welcoming and hearing them, patients may come to understand they share common experiences with other patients, which may help them to feel less emotionally disorientated. Secondly, the presence of a peer beside them can also encourage testimony on the specific affects they are experiencing and the reasons for it. For instance, if the patient wishes to talk about the anger they feel about their diagnosis, the presence of another patient may help and encourage them in an explicit ("go on") or non-verbal ways (such as by nodding their head), showing them there is an empathetic listener. Thanks to this resource person, patients may be less likely to self-silence their affects, whether in a one-to-one relationship with the patient partner/advocate or in the presence of a healthcare provider.

However, this possibility does not clarify to what extent the expression of these affects may improve care. While there is value in having affects stay in a confidential relationship between peers (as seen in a previous chapter on narration), in many cases affects may say something important about healthcare decision-making. As I have already elaborated in the previous discussion, while healthcare providers should provide the epistemic and affective safe space for the patient, this is not always the case. While epistemic authority may be principally the cause, patients also self-silence their affects by no fault of their healthcare provider: due to fear of not being taking seriously, or to be seen as "too emotional," patients may not feel comfortable discussing their affects in front of their doctors. This may happen even in cases in which the healthcare provider has the best intentions and actively solicits the patient's affects.

While there is not enough literature on what changes when a patient partner/advocate is integrated in a discussion on shared decision-making, I can see several possibilities. First of all, when this person is integrated into a discussion, the relationship changes, as there are now two patients in front of a doctor. This person is also one may understand what they are going through and the emotions they are experiencing. While the healthcare provider should already encourage this kind of safe space for the patient, it may be easier for the patient when there is someone beside them who fully understands and encourages them to speak up. When this takes place in the context of shared decision-making or another moment of interaction between the healthcare provider and the patient such as in therapeutic patient education, the healthcare team may have access to new information or experiences—including affects—that would not have been made possible without their presence.

Furthermore, I propose that having a third person in the room may also foster the kind of critical self-awareness needed for healthcare providers to be made aware of how they interact with patients. Having the patient partner/advocate present makes the consultation an unusual one, one in which the healthcare provider can observe themselves and the interactions with their patients. It may therefore create a space in which they take more time to encourage the patient's testimony.[3]

I am not advocating here that the person should be integrated into every kind of consultation. This is likely beyond the means of the patient partner/advocate (who largely exercises in a voluntary capacity) and it may not be helpful in all temporalities or spaces. However, if the patient desires the presence of this resource person, in particular if it helps them to feel more comfortable expressing their experiences, worries, and fears, it could be a useful resource to bring out the patient's affects and have them be heard. It may be particularly helpful for those with vulnerabilities who have a hard time participating in the consultation—or being heard and understood by healthcare providers.

Furthermore, this affective opportunity is not only confined to the direct relationship between a patient/healthcare provider/patient partner. If the patient partner/advocate is integrated into interdisciplinary meetings or other forums of collective discussion, this may also give an opportunity for their affects to be heard and understood even when the patient is no longer present. For instance, by integrating an interdisciplinary discussion, they may be able to relate information they have gathered in their freer discussions with the patient. This condition should of course only be made possible if the patient agrees that this information be shared; however, when patients find themselves self-silencing their affects in front of their healthcare

[3] While this is a hypothesis on my part and merits further research, I can already witness the effect of a third person presence when I have participated as an observer in a consultation. Having a third person in the room means taking extra time, paying attention to the issues that have been highlighted during the discussion, etc. There is thus good reason to believe the presence of another patient can help doctors to practice this kind of critical self-awareness and take time for issues that may be glossed over in other moments.

provider, they may find the ability to relate this information by another means or person useful. I am thinking here in particular of when the patient does not feel comfortable expressing embarrassing or difficult to express affects in front of their healthcare provider (such as suicidal thoughts, depression, or anxiety), but which may be important to the treatment plan and patient safety.

However, this proposition brings forth the question of whether or not the patient partner/advocate: (1) should be involved in interdisciplinary care meetings alongside healthcare staff; (2) if we can consider them care workers, and if yes, how to hold them to the same standards as healthcare providers in guaranteeing patient privacy and informed consent, including in communicating affective information to other healthcare providers. These issues are some of the reasons their integration remains a contentious issue and their inclusion resisted. While these issues remain to be solved and their roles clarified, for our discussion of affective justice, their inclusion in interdisciplinary discussions appears promising. It is promising as it gives a forum for the patient's affects to be shared and recognized among the team of healthcare providers. However, if we do decide to integrate them in these spaces, it is clear that training and codes of conduct are needed for patient partners/advocates to be aware of and respect patient's wishes, including in the divulgation of their affects to other healthcare providers.

As we have seen, despite ongoing resistances to their inclusion, patient partners/advocates have the potential to help work toward affective (and epistemic) justice for patients. This includes providing safe spaces for patients because they have found someone who understands them, but also having this knowledge be better heard and understood outside of these spaces and among the healthcare team. Given that affects may very well affect quality care, their inclusion is therefore not to be discounted as a way to work toward greater affective justice.

Can patient partners/advocates also be sources of affective injustice?

This however brings us to the last question in this debate, namely if these resource persons may also become the source of affective (and epistemic) injustice for patients. While I see many reasons why they may provide the kinds of affective resources, opportunities, and recognition we are seeking, there is also the risk in integrating a third person into the conversation.

First of all, there is the risk when the patient partner/advocate does not hear or understand the patient they are supposed to be helping. This kind of situation may happen because their own experiences do not resonate with the patient they are supposed to be helping, or because they do not understand or hear them for various reasons. While it is supposed that these persons are closer to patients, each person is an individual with a unique history and experience of illness. Their experiences may not always converge.

The first way affective injustice may occur is in conversation with the patient. For instance, if a patient says they are scared for a risky medical procedure and do not wish to go ahead with it, the patient partner/advocate may interpret this fear as irrational. They may consider it so because they have themselves gone through the procedure and it went well, because they may have forgotten they were fearful, or because they have since learned more about such procedures by being integrated into the healthcare team. This may lead them in some cases not to take the affect seriously or to dismiss it, saying it is irrational or misplaced, rather than providing the safe space in which these fears can be openly discussed and recognized by their peer.

Secondly, patient partners/advocates are also susceptible to bias and discrimination toward certain groups—just like healthcare providers, just like society at large. They may unfairly judge or misunderstand the patient based upon a prejudice. This negative evaluation may be geared toward their affects, such as the idea that a person from marginalized group is more "emotional" or they "exaggerate" their fear or worry about their illness or treatment. Bias may also come from the social status of the patient partner/advocate. A countrywide study in Canada found that most persons self-identifying as patient partners were retired white females over the age of 50 with a high socio-economic status (Abelson et al. 2022). Similar patterns of unequal representation in patient partnerships have been frequently discussed elsewhere, raising questions on who these persons represent (Dukhanin et al. 2020). I have been advocating throughout this book the need to cultivate critical self-awareness for everyone (whether a patient partner, healthcare provider, or family member) to become aware of their bias and to practice affective humility toward others. This also includes a recognition of the bias that patient partners/advocates may also have and develop toward certain social groups (because of privilege, or unconscious or conscious bias and prejudice) that may be perpetuated in their interactions with patients.

In turn, when patient partners/advocates come to function as "reporters of patient experiences," notably because they have been trained or learned to reformulate certain experiences—in words and language more acceptable or understandable to the healthcare team—the risk also becomes that they badly formulate or misunderstand these affects and what they might mean. This situation may also occur when they use the patient's confidences to make an evaluation about the person's affects. The first risk is this person will no longer be an affective resource—but rather cause epistemic and affective harm to the patient. Another risk is when the healthcare team believes the patient partner/advocate—over the patient and their family—because they seem to incarnate an epistemic authority about the illness experience compared to the patient. All of these situations may turn into affective injustices stemming from patient partner/advocate involvement.

As seen from these examples, while patients share many things in common in their experiences of a chronic illness, they are also individuals, with different approaches, beliefs, affects, and experiences. As the discussion on affective humility in Chapter 3 has shown, there is a need to recognize that we will never fully have access to the other's affective experiences, although they may in some instances closely resemble our own. Therefore, patient partners/advocates will also need to cultivate epistemic and affective humility in order to serve as affective resources for other patients.

Given these risks, let us consider what we can do so that these persons can fully benefit patients and the healthcare team. First of all, we can partly guard against this risk by prioritizing the patient and what they desire and the means they wish to express it. This means the first priority will be to invest in the healthcare provider-patient relationship and provide a safe affective space through the singular colloquy. Another way to guard against this risk is to invest in clear communication to the patient of their right to include or exclude the patient partner/advocate. This person must also clearly solicit the informed consent of the patient, which includes clearly stating the information they wish to provide to the healthcare team. Finally, there is a need to invest in training for them—as for other members of the healthcare team—to help them recognize their own biases so that they can practice both humility and critical self-awareness of how they treat persons in function of their social status.

To conclude this section, these issues suggest patient partners/advocates have the potential to serve as affective resources for patients and their families, if done carefully. However, careful attention must still be paid to individual patients and their experiences and affects. In other words, patient partners/advocates may provide the bridge to communication with healthcare providers, but they cannot replace them. The singular colloquy between the healthcare provider and the patient is not to be forgotten, even while new forms of citizen engagement are moving forward.

6.2.3 *Patient Associations: A Safe Space for Affects?*

In our final discussion, I will look at the role of patient associations to provide a safe space outside of the hospital for patients to express their affects. As in the previous discussion, I will discuss the potential—but also the risk—of thinking of them as affective resources and opportunities.

To begin with, while patient associations have been active since at least the 1960s, their role and importance has been notably recognized from the 1990s, with the important role they played in the AIDS crisis. While their roles vary according to the specific context, financing, and illness, patient associations today fulfill a variety of advocacy and support roles for patients. This includes lobbying for changes in healthcare policy, gaining access to medication and treatment, but also participating as patient representatives in ethical committees, research projects, and policy forums. In their day-to-day work, they also provide peer support through individual and group assistance to their members. This emerging role often includes a mission to help reduce the isolation of their members with social activities and provision of support for family members. As seen from this description, not only are they participating in most forms of health democracy, but they also have a vital role in providing individual peer support to their members.

As we will see from this discussion, patient associations provide similar affective resources and opportunities as patient partners/advocates in the hospital. In this section, I will in particular discuss the provision of peer support, as this is the most realistic way in which affective justice may be possible for patients. However, it is of course to be acknowledged that serving as patient representatives in other ways such as lobbying and advocacy has both direct and indirect effects on patients, as it means prioritizing certain issues about an illness compared to others. While advocacy and policy change may benefit the overall patient population, it may also cause harms to those who are outside of the dominant hermeneutical understandings of the illness. This is an important issue of epistemic injustice that merits further exploration; however, as it is a separate issue to our debate in this chapter, I will not treat it here.

In terms of provision of peer support through patient associations, this can take place in a one-to-one conversation between the patient representative and the person affected by the illness, who may need counselling, support, or advice. However, it may also happen in group discussions, such as monthly meetings in which persons are invited to speak about their experiences. In both of these methods, affects are likely to play a central role, as the conversations will be about illness experiences. In them, there is an important listening and recognition role for those around the table. There is also a vital role for the group facilitator, who is there to guarantee a safe space for all members present.

To better understand the affective opportunities offered by peer support, I will take a case study from my research of monthly epilepsy patient association meetings in France, in which I conducted ethnography for over a year in two separate association meetings, in two cities (Lyon and Grenoble). These meetings were held once a month and all patients with epilepsy were invited. One was run by the main patient association in France (Epilepsy France) while another was an independent group created by a patient, who wished to provide a different and independent space for discussion. However, both used a group discussion format, in which persons were free to come with their experiences, questions, and worries. Given the social isolation experienced by persons with this illness, a secondary goal was to provide a space for social interaction.

What I witnessed was that these group discussions provide a variety of functions based upon the needs of their group members. Persons around the table are invited to ask questions, share advice, or simply to discuss what is on their mind. They therefore function to provide information, access, and contacts to patients newly diagnosed with epilepsy, helping them to navigate the often-opaque healthcare journey from generalist to specialist. They also help them to understand medications, treatments, and the realities of their illness in the social context. However, given their free format, they also function to "let out steam," meaning that they are forums in which complaints, fears, uncertainties, and anger can be freely expressed. While these groups rarely make affects an explicit focus of their meetings, they are an inevitable part of them. Furthermore, what I witnessed is that these affects were taken seriously by their fellow patients, even in the case of negative emotions such as fear or anger, as patients could relate to and often had similar experiences and feelings about living with the illness in society (Bogaert 2020).

One example that marked me in particular was listening to patients talk about suicide. Patients frequently evoke their fear of seizures and the stigma and discrimination this leads to in their interaction with others. This happens in particular in their workplaces, where they are put out of the way or given the dirty jobs no one wants as they are seen as a "risk" due to their seizures. The social isolation they feel in the workplace and in other areas of their lives often leads to feelings of anxiety, sadness and depression. For some, this may also lead to suicidal feelings—and unfortunately for some—an attempt to end their lives. Although patients had a hard time discussing these subjects with their healthcare providers, and with their family and friends, in the patient group they often found a safe space. What was striking to me was that they were able to discuss these difficult experiences extremely openly. This was because their peers around the table had lived through the same experiences, the same periods of depression, the same fears and anxieties, the same suicidal thoughts. Their affects were heard by the others around the table. They were taken seriously and understood.

However, the possible danger in this type of discussion was that it could lead to a sort of "emotional contagion" in which participants might leave feeling more depressed and anxious than when they came to the meeting in the first place, as such types of interactions bring up difficult emotions. While it is important to acknowledge this is a possibility (and here fellow patients and group leaders can play an important role by paying attention to these situations and referring them to further support if needed), what I personally have witnessed is that even though these experiences were heavy, they also made patients feel less alone and even comforted afterwards. This was because others' experiences gave a certain legitimacy to the affects they experienced. If we mobilize the terminology from the affective injustice literature, we can say that it not only provided a forum for them to be heard (affect-related testimonial justice) but also to be understood (affect-related hermeneutical justice). This was possible as they learned they were part of a collective illness experience and not "weird" or "isolated," which they may have continued feeling if they did not have this group resource.

In terms of improving care delivery, these persons are outside of the hospital, so it is unlikely they would have a direct impact by providing this kind of affective resource. However, in an indirect way, they did serve as epistemic resources for other patients in ways that may improve care and well-being for the individual. In the discussions I witnessed about suicide, peers: (1) frequently alerted other patients to possible side effects of certain medications, including feelings of sadness and depression that could be causing their suicidal thoughts; (2) shared resources on neurologists, psychologists, or other support persons to help them if needed if they felt in crisis. In this way, they did make the link between the hospital and the patient, serving as both affective and epistemic resources for them and likely improving care delivery and the person's well-being.

Peer support in this way plays an important role for patients in having their affects be heard and understood and in reducing their emotional disorientation. While this does not solve the problem of whether these affects may be recognized in outside forums, it is a realistic means of working toward affective justice for patients (at least in these spaces). This does not put doctors or institutions "off the hook," as there is still much work to be done in the hospital space for epistemic and affective justice to be achieved. However, when looked upon as a complementary resource, they do show considerable promise to work toward affective justice for these patients.

Can patient associations be a source of affective injustice?

While the discussion thus far has shown the benefit of peer support in providing affective safe spaces, I now come to the more uncomfortable topic, in that affective injustice may also play out in these groups themselves. As in our previous discussion, while patients share common experiences of illness, not all are equal in terms of social status (age, profession, educational background, etc.). They are also not equal in terms of their roles in patient associations, with those serving as their representatives/leaders taking on a certain epistemic authority in these types of meetings. Because of these hierarchies, it is to be expected that epistemic injustices—and affective injustices—will also play out in these groups due to these differences. Patient association meetings are often led by persons with higher educational backgrounds, who are more comfortable leading and organizing advocacy campaigns, communicating with healthcare specialists, or being the face of the media. This means that they are likely to take this role as well in deciding what information, what affects, what forms of patient experiences are legitimate. In my own experiences of observing patient association meetings, one group leader was a patient who worked in research; in the other patient group, the leaders were family members or support workers (social workers or doctors) that had an interest in defending patient rights. In both instances, therefore, the leaders came from privileged groups and backgrounds.

The group members themselves however often came from more disadvantaged backgrounds, at least in the sense that they often did not achieve a high level of education or high-status profession due to their illness and the effects of treatment. Due to memory issues and the effects of epilepsy medication, which often cause confusion and tiredness, it is difficult for some persons to achieve the same level of educational or professional achievement as persons who do not face this particular illness. Because of these vulnerabilities, most members would have had trouble taking leadership positions in strategic committees or in advocacy with their doctors; it is perhaps understandable that those taking leadership positions therefore came from these more privileged groups. However, this created a certain inequality in knowledge recognition. While these group leaders showed a critical self-awareness and tried to cultivate each group member's testimony, in some cases they favored some testimony or members over others. This was often unintentional, but it is a reality of group dynamics in patient groups and the epistemic authority of those who take leadership roles.

Taking these considerations into mind, if we advocate that these groups be seen as a privileged space for patients in terms of affective justice, we can also see the limits and the special attention that must be taken to guarantee the epistemic and affective contributions of all members. This includes paying particular attention to those members who may be excluded or self-silenced. We must also pay attention to the fact that these safe spaces are outside of the hospital space—and we have not addressed affective injustice taking place within hospital structures and in interactions with healthcare providers, who (as I have shown) may have a harder time dealing with, responding to, or understanding these affects. We therefore conclude this section with a better understanding of the ways that patient associations can serve as prominent affective resources and opportunities, but also with greater awareness of the caution that needs to be taken to ensure they function as safe spaces for all patients.

6.3 Patient Engagement and Healthcare Provider's Affects

I would now like to explore to what extent patient engagement may also be beneficial to healthcare providers. So far, this chapter has largely focused on the patient's affective needs through greater patient engagement. However, can patient engagement improve the expression of healthcare provider's affects? While this concept seems less promising than our discussions on patients, I do see a number of ways that the inclusion of the patient partners/advocates can also change how healthcare is thought and delivered, thereby paving the way toward greater affective resources and opportunities for healthcare providers.

First of all, patient engagement works toward an ethics of care, one that is person-centered. The presence of a patient/partner advocate in this way decenters the discussion away from the values of efficiency and efficacy as the only ones relevant. They pave the way toward a greater care space based on the values and needs of patients, including their affective needs. In this way, they contribute to a culture change in the hospital and provide the space for other healthcare values—be it narration, spirituality, or emotions—to come to the forefront. Thus, while their integration may not seem relevant at first to valuing healthcare provider's affects, by promoting a change in ways of caring, they may work toward affective justice for healthcare providers as well.

In the second place, the involvement of these resource persons in the hospital service—or the involvement of patient representatives in other forums—also provides an outside/neutral person with whom the healthcare provider may feel more comfortable in expressing their affects. For instance, while a nurse may not feel comfortable talking to their fellow nurse or the doctor about the bad day they are having, they may let off steam, laugh, or talk in an informal way with the patient partner/advocate with whom they may have a more comfortable, informal, or non-hierarchical relationship. Like the hospital chaplain seen in Chapter 5, providing a non-healthcare provider to the care team provides the kind of informal sharing that may be conducive to the provision of informal affective safe spaces. The extent to which this is possible will of course depend on their relationship, the functioning of the team, and how the healthcare provider recognizes or sees them. It is however one possibility to work toward a change in the hospital, one in which affects may be seen and recognized.

Can patient partners/advocates be the cause of affective injustice for healthcare providers?

However, there is—like in our discussion of patients—also a risk that needs to be taken seriously for the healthcare provider. While the jury is still out in whether or not patient partners/advocates should be included into team discussions, their involvement could represent a risk in terms of affective justice. As I have shown in Chapter 4, interdisciplinary meetings often function as affective safe spaces, allowing healthcare providers to blow off steam by speaking frankly and directly about experiences of patient care. They can do so because they are among fellow colleagues and know that this "dark humor" and other forms of letting off steam will not be heard by the patient or their family member. Thus, while the patient partner/advocate can help work toward affective justice for the patient by participating in these meetings, they may have the opposite effect on healthcare professionals, who may choose to silence their affects in front of the patient partner/advocate. This will happen in particular in instances in which the person is not seen or recognized as a team member, but as a patient who may be negatively affected by what they say and the way they say it.

Secondly, it can be the case that the patient partner/advocate may silence the affects of the healthcare provider. If they serve as the epistemic authority about the patient, this could mean the healthcare provider's feeling toward and experiences with the patient are devalued, as they "don't understand" as they are not a peer. While this is not a frequent occurrence at the current time given resistances to their inclusion, it could be in coming years as their involvement becomes more formalized. While I have argued throughout this chapter in what ways their engagement may help improve care, it is a risk that merits careful consideration. As we can see, in these discussions of affective justice, we will need to think at multiple levels and understand how to provide a safe affective space for all, whether the person is a healthcare provider, a patient, a family member, or a patient partner/advocate. This is a delicate balance, but one that can be navigated by fostering the critical self-awareness of everyone involved.

6.4 Affective Injustice Experienced by Patient Partners/advocates

In this final section, it will be important to pay attention to the ways in which patient partners/advocates may also experience affective injustice whilst trying to work with healthcare providers and institutions and what we can do about it.

First of all, there still exists an asymmetry between patients and healthcare providers, and the ways in which patients may still be subject to a deflated level of credibility due to their social status as patients. This hierarchy also plays out in the engagement of patient partners/advocates. Concretely, this means that their knowledge—including their affective knowledge—is not always heard or understood because of their social status. This can lead to affective injustice for these resource persons, particularly when their affective experiences are different from healthcare providers' or in instances in which their own affects are not considered relevant to the discussion. This is a serious problem, as affective experiences are an important part of the experiential knowledge that the patient partner/advocate can bring to the table.

Furthermore, as there is a push to professionalize these persons, there is also an expectation that they remain affectively neutral in order to work together in the healthcare team. By affectively neutral, I mean an expectation that these persons cannot react or express themselves emotionally in a team. Given the asymmetry of this relationship, as well as conceptions that patients will be "more emotional" than healthcare professionals due to their illness experience, there may be implicit and explicit pressure by the healthcare team—and by patient partners/advocates themselves—to silence their affects to justify their place in the team.

Finally, not being recognized or understood, or being forced to silence their affects, may have a detrimental impact on the well-being of these persons. These persons can be patients, former patients, or family members. Many of them have experienced traumatic or difficult experiences during their healthcare journey, in particular because of the kind of asymmetries that we have been discussing. These difficult experiences—and explicit and implicit attempts at self-silencing—may have a negative impact on them, as well as their motivation and desires for engagement. It may also increase or cause their emotional disorientation about their illness experience. Therefore, while these persons may be increasingly seen as professionals, we need to make sure their affective needs are also not ignored.

To better understand these issues, Richards et al.'s (2023) experiences of patient engagement in research projects is particularly insightful. The authors are six patient partners in Canada who have participated in a variety of successful and unsuccessful types of patient engagement over the years. Their article aptly named, "Reflections on patient engagement by patient partners: how it can go wrong," explores the often symbolic roles they take in patient engagement forums and the epistemic and affective injustices they often experience. Examples of what can go wrong include being recruited as co-chairs of meetings but without full support to create or influence the agenda, or having feedback and contributions minimized or dismissed when it does not fit the dominant agenda. In their article, they also highlighted where they were excluded from engagement based on age, ability, race, diagnosis, or other components of an intersectional identity.

Richards et al. highlight many areas of epistemic and affective injustice. They show that patients are often invited to meetings in order to agree with what is said (therefore fulfilling a symbolic role). Even worse however is when they bring up thoughts or beliefs that disagree with dominant opinions, which are discouraged, not heard or understood. They also relate experiences of collaboration where they were told that their perspectives are "biased" and "emotional" (thus irrelevant). This occurred in particular when professionals on the team do not share the same opinion. These are clearly examples of affective and epistemic injustice, including silencing of their knowledge and affects.

According to these authors, those experiences that "went wrong" leave patient partners feeling gaslit[4] and questioning the validity of their affects, especially by encouraging the perception they are being overly sensitive or non-objective for expressing them. Furthermore, as they discuss, patient partners bring experience and expertise which are sometimes very painful or traumatic. The expectation they should remain "objective" and "non-emotional" can therefore lead to an even more traumatizing and alienating experience for them, particularly when there is no safe

[4] By this term the authors are referring to a form of manipulation that causes the victim to doubt their thoughts, knowledge, or reality, which can also lead to a loss of confidence or of their emotional stability. For a historical overview of the term, see: Klein and Bartz (2023).

space created for them to discuss these difficulties during their engagement. Careful attention therefore needs to be paid to the affective needs of patient partners/advocates for such types of collaboration to be successful and to avoid further harm.

6.5 Conclusion

This chapter has explored the potential and the limits of fellow citizens to serve as affective resources and opportunities. As this issue develops in the coming years, we will better understand in what means, temporalities, and spaces citizens can best serve the affective needs of patients and healthcare providers. However, this chapter has already given some preliminary ideas about the ways they may create a safe space for affective (and also epistemic) expression. As our discussion has shown, their place remains fragile and we will need to overcome resistances to their inclusion to fully benefit from their experience and expertise. It will also be important in the coming years not to ignore the possibility of an increasing delegation of affective tasks to patient partners/advocates, who as we have seen, are often viewed as the "emotional" ones due to their experiences of illness. As we saw in the last chapter (spirituality), the temptation to delegate all those parts of patient experience not related to biomedical diagnosis and treatment to other persons is recurrent. If not the psychologist or psychiatrist, then the hospital chaplain, and potentially now the patient partner/advocate may be delegated to take care of the patient's affects. While patients may have specific tools, experiences, and life perspectives that enable new ways for affects to be valued, that does not mean they are always the right affective resource for that particular patient, or their contribution should be limited to this type of role. The whole healthcare team, as I have shown throughout this book, will need to provide the affective resources and opportunities to help the patient receive better care. The risk of this delegation is also if the patient partner/advocate is the source of the affective (or epistemic injustice), then they may not have any resources persons in which to turn.

References

Abelson, Julia, Carolyn Canfield, Myles Leslie, Mary Anne Levasseur, Paula Rowland, Laura Tripp, Meredith Vanstone, et al. 2022. "Understanding Patient Partnership in Health Systems: Lessons from the Canadian Patient Partner Survey." *BMJ Open* 12 (9): e061465. https://doi.org/10.1136/bmjopen-2022-061465

Ali, Khalid, Christine Roffe, and Peter Crome. 2006. "What Patients Want : Consumer Involvement in the Design of a Randomized Controlled Trial of Routine Oxygen Supplementation After Acute Stroke." *Stroke* 37 (3): 865–871. https://doi.org/10.1161/01.STR.0000204053.36966.80

Arveiller, Jean-Paul, and Philippe Tizon. 2016. "Démocratie sanitaire, qu'est-ce à dire ?" *Pratiques en sante mentale* 62e année (2): 2–2.

Bogaert, Brenda. 2020. "Proposition méthodologique pour une philosophie de terrain avec les émotions à la lumière de la pensée de Martha Nussbaum." Application/xhtml+xml,application/pdf,online,database,application/xhtml+xml ; application/pdf, 18 pages, pages 75–92. https://doi.org/10.15122/ISBN.978-2-406-10144-4.P.0075

Bogaert, Brenda. 2023. "Les Patients Partenaires Dans Des Recherches En Santé : Les Enjeux Éthiques et Épistémologiques à Prendre En Compte Pour Concevoir Une Collaboration Fructueuse:" *Revue Française d'éthique Appliquée* N° 13 (2): 131–142. https://doi.org/10.3917/rfeap.013.0131

Bombard, Yvonne, G. Ross Baker, Elaina Orlando, Carol Fancott, Pooja Bhatia, Selina Casalino, Kanecy Onate, Jean-Louis Denis, and Marie-Pascale Pomey. 2018. "Engaging Patients to Improve Quality of Care: A Systematic Review." *Implementation Science* 13 (1): 98. https://doi.org/10.1186/s13012-018-0784-z

Boote, Jonathan, Wendy Baird, and Claire Beecroft. 2010. "Public Involvement at the Design Stage of Primary Health Research: A Narrative Review of Case Examples." *Health Policy* 95 (1): 10–23. https://doi.org/10.1016/j.healthpol.2009.11.007

Borkman, Thomasina. 1976. "Experiential Knowledge: A New Concept for the Analysis of Self-Help Groups." *Social Service Review*, September 1976, University of Chicago Press edition, sec. Vol. 50, no. 3.

Crogan, Neva. 2021. "What Happened to Person-Centered Care? Observations During the COVID-19 Pandemic." *The Journal for Nurse Practitioners: JNP* 17 (3): 367. https://doi.org/10.1016/j.nurpra.2020.11.001

Dogba, Maman Joyce, Matthew Menear, Dawn Stacey, Nathalie Brière, and France Légaré. 2016. "The Evolution of an Interprofessional Shared Decision-Making Research Program: Reflective Case Study of an Emerging Paradigm." *International Journal of Integrated Care* 16 (3): 4. https://doi.org/10.5334/ijic.2212

Dukhanin, Vadim, Scott Feeser, Scott A. Berkowitz, and Matthew DeCamp. 2020. "Who Represents Me? A Patient-derived Model of Patient Engagement via Patient and Family Advisory Councils (PFACs)." *Health Expectations* 23 (1): 148–158. https://doi.org/10.1111/hex.12983

Elwyn, Glyn, Dominick Frosch, Richard Thomson, Natalie Joseph-Williams, Amy Lloyd, Paul Kinnersley, Emma Cording, et al. 2012. "Shared Decision Making: A Model for Clinical Practice." *Journal of General Internal Medicine* 27 (10): 1361–1367. https://doi.org/10.1007/s11606-012-2077-6

Gallegos, Francisco. 2021. "Affective Injustice and Fundamental Affective Goods." *Journal of Social Philosophy* n/a (n/a). https://doi.org/10.1111/josp.12428

Higgins, Tracy, Elaine Larson, and Rebecca Schnall. 2017. "Unraveling the Meaning of Patient Engagement: A Concept Analysis." *Patient Education and Counseling* 100 (1): 30–36. https://doi.org/10.1016/j.pec.2016.09.002

Klein, Willis, and Jennifer A Bartz. 2023. A Historical Review of Gaslighting: Tracing Changing Conceptualizations Within Psychology and Psychiatry. Unpublished. https://doi.org/10.13140/RG.2.2.30013.23527

Manias, Elizabeth, Carmel Hughes, Robyn Woodward-Kron, Guncag Ozavci, Christine Jorm, and Tracey Bucknall. 2024. "Decision-Making about Changing Medications across Transitions of Care: Opportunities for Enhanced Patient and Family Engagement." *Research in Social and Administrative Pharmacy*, February, S1551741124000494. https://doi.org/10.1016/j.sapharm.2024.02.002

Pierron, Jean-Philippe. 2006. "L'engagement. Envies d'agir, raisons d'agir." *Sens-Dessous* N° 0 (1): 51–61.

Richards, Dawn P., Sabrina Poirier, Vina Mohabir, Laurie Proulx, Sue Robins, and Jeffery Smith. 2023. "Reflections on Patient Engagement by Patient Partners: How It Can Go Wrong." *Research Involvement and Engagement* 9 (1): 41. https://doi.org/10.1186/s40900-023-00454-1

Skovlund, Pernille Christiansen, Berit Kjærside Nielsen, Henriette Vind Thaysen, Henrik Schmidt, Arnstein Finset, Kristian Ahm Hansen, and Kirsten Lomborg. 2020. "The Impact of Patient Involvement in Research: A Case Study of the Planning, Conduct and Dissemination of a

Clinical, Controlled Trial." *Research Involvement and Engagement* 6 (1): 43. https://doi.org/10.1186/s40900-020-00214-5

Thomas, Aliki, Ayelet Kuper, Benjamin Chin-Yee, and Melissa Park. 2020. "What Is 'Shared' in Shared Decision-making? Philosophical Perspectives, Epistemic Justice, and Implications for Health Professions Education." *Journal of Evaluation in Clinical Practice* 26 (2): 409–418. https://doi.org/10.1111/jep.13370

Vanstone, Meredith, Carolyn Canfield, Cara Evans, Myles Leslie, Mary Anne Levasseur, Maggie MacNeil, Manisha Pahwa, et al. 2023. "Towards Conceptualizing Patients as Partners in Health Systems: A Systematic Review and Descriptive Synthesis." *Health Research Policy and Systems* 21 (1): 12. https://doi.org/10.1186/s12961-022-00954-8

Chapter 7
Affective Resources and Opportunities via Hospital Spaces and Architectures

Abstract In this chapter, I will look at the affective opportunities and resources for patients, families, and healthcare providers that may be facilitated by hospital design and architecture. Building upon the affective injustice discussion, theories on emotional design and architecture, and research in evidence-based design, I will suggest several ways that hospital design and architecture may improve well-being. In the first discussion, I will explore how hospital spaces, and in particular the hospital waiting area and the patient room, can facilitate greater emotional security. In the second, I will elaborate the room environment, notably the influence of light, color, and sound on the user's affective experience. The third subject, inspired by architectural theorist Giuliana Bruno's theory of e(motion), will explore the hospital space as *vissuto*, a space of lived experiences that invites us to rethink design and architecture to provide for greater user interaction.

Keywords Emotions · Light · Color · Hospital spaces · Hospital architecture · Hospital design

7.1 Introduction

In this chapter, I will be interested in what affective resources and opportunities may be provided through hospital design and architecture. I will use Gallegos' (2021) definition of resources and opportunities to mean those activities, spaces, and circumstances that may contribute positively to one's subjective well-being. As we will see, hospital design and architecture can both contribute positively to well-being as well as help relieve emotional distress in hospital users.

My main question is how can a hospital design or architecture be healing for its users. In the first place, the design of space in a hospital[1] can favor a culture,

[1] Parts of this article were first published in the journal *HERD: Health Environments Research & Design Journal* and have been reused here with the permission of the journal. However, the discussion in this chapter has been substantially expanded, updated, and adapted for the purposes of the affective injustice discussion. For the original article, see Bogaert (2021).

B. Bogaert, *Affective Injustice in Healthcare*, The International Library of Bioethics 112, https://doi.org/10.1007/978-3-031-94378-2_7

complementary to or in resistance to a technical culture that can change the experience lived by the user for which the project is designed (Pellerin and Coirié 2017).By paying attention to spaces and architectures in hospitals, care can become more than just a cure, by permitting persons to take care of themselves and others in a hospital environment (Pierron 2019). In most evaluations today, these kinds of intangible values, which are hard to capture and measure in hospital evaluations, have been sidelined in favor of indicators such as efficacy and efficiency (Pierron and Vinot, 2020). However, increasing evidence suggests that design and architecture can make significant differences in patient satisfaction, quality of life, lowering pain medication use, and recovery times, among other care outcomes (Lawson 2010; MacAllister et al. 2019). Hospital design and architecture is therefore to be taken seriously from a public health, management, person-centered, and affective perspective.

In this chapter, I will cite evidence-based studies in environmental design that will enable me to show how certain aspects of design (be it the hospital bed, color, or light in the room) may contribute positively to the person's subjective well-being. While the hospital space is often designed for functional needs (efficiency, safety, and decreasing medical errors), designers are increasingly recognizing the importance of taking into account how to create *healing spaces* by considering other needs, including affective needs (DuBose et al. 2018).

In terms of my choice to use evidence-based design to develop this subject, while there is increasing evidence to support the link between design and well-being, it must be recognized that high quality random control trials are still needed to validate this evidence and to be able to generalize across patient populations (Dijkstra et al. 2006). In addition, while I will focus on individual aspects of design and architecture in order to see the potential of each object/design strategy, it is to be taken into consideration that each strategy can only be understood as part of a whole. For instance, a hospital waiting room that has a plant or uses green imagery cannot provide a calming environment by itself; it must be accompanied by comfortable chairs, colors that calm, and attention to the overall sound environment (soundscape). Furthermore, a room with a view outside to nature, but one in which the bed is uncomfortable or is perceived as unsafe, will not make it a safe affective space. In other words, hospital design cannot become an affective resource for the patient, family, and healthcare provider without considering how the user will feel about the hospital space and experience as a whole.

Furthermore, as I have been advocating throughout this book, it is not enough to provide a promising affective resource such as the ones I will describe in this chapter alone. They will need to be combined with other resources to provide an environment in which affects are taken seriously and recognized. Therefore, while I will advocate for an attention to design and architecture, we should also not forget that the care experience—and the affects that are lived and shared in the human relationships of care—are the heart of the matter. A calming hospital environment will do little to make the care experience more hospitable if patients are left alone too long in the hospital space, if they have no say in the decisions taken, or if their relationships with their healthcare providers are conflictual. However, as I will show, hospital design and architecture can contribute to making the care experience and working environment a more hospitable one when integrated into other factors of person-centered care delivery.

In this discussion, I will use Ulrich's theory of supportive design (1991) to help to articulate how hospital space and design can function as an affective resource. Ulrich conceptualizes the ways in which the healthcare environment can reduce stress and promote well-being by responding to three needs: (1) a sense of control; (2) access to social support; (3) access to positive distractions. We will see how these three needs play out in this chapter through various design strategies to help hospital spaces become affective safe spaces for patients. Table 7.1 summarizes the strategies that I will discuss that have the potential to provide greater well-being for patients, healthcare providers, and families. While no means an exhaustive list, I have chosen them for discussion as they are the most researched and used at the current time in hospital architecture and design.

In order to prepare the discussion, I will first give an overview of how architects and designers have theorized and understood the articulation of affects in their work. I will then move on to practical case studies in hospital design and architecture, in particular through examples from evidence-based design. My first case study will elaborate how the design of a hospital space may encourage emotional security and a sense of control (Ulrich's first level), and to illustrate this, I will focus on the design of the waiting room and the hospital room. I will then show how paying attention to light, sound, and color may also influence the patient's and family's affective experiences (Ulrich's first and third levels). Finally, taking into account architectural theorist Bruno's reflection on *vissuto* (2011), I will see how we can redesign hospital spaces for interaction and participation (Ulrich's second level).

Table 7.1 Providing affective resources through hospital design and architecture

Strategies	**Ulrich's supportive design** First level: sense of control Second level: social support Third level: positive distraction
1. Use of nature (direct and indirect)	Third level
2. Individualized spaces of interaction and privacy	First/second level
3. Design that promotes safety, security, and comfort	First level
4. Use of natural light/access to windows	Third level
5. Attention to the soundscape	First/third level
6. Colors that calm and help navigate the hospital space	First/third level
7. Spaces for activity, interaction, and rest	First/second/third level

7.2 Emotions in Theories of Architecture and Design

Before getting started with our case studies, it is worthwhile taking some time to detail how architects and designers have understood the importance of health, safety, and welfare as part of their mission, including designing spaces and architectures that engender positive emotional responses from their users (Dannenberg and Burpee 2018). In this section, I will briefly explore the diverse movements and theorists that have conceived of the interrelation of affects and design.

First of all, it must be said that the recognition of the importance of affects to architecture and design has long been prioritized. Emotion in architecture was already documented from the eighteenth century with French architectural theorists such as Nicolas Le Camus de Mézières (1721–1773). This theorist believed that buildings could evoke sensations in the person, as they can speak to the mind and move the soul. In a similar way, Etienne-Louis Boullée (1728–1799) advocated that architecture can communicate moral principles by modulating the lives and emotions of its inhabitants (Pelletier 2008, p. 2). The emotional architecture movement, instigated by Mexican architect Luis Barragán and the sculptor Mathias Goéritz's 1953 *Emotional Architecture Manifesto*, continues this tradition. They claim that the modern man has been crushed by an excess of functionalism. Arguing for an architecture that can move its inhabitants, their creations seek to stimulate positive emotion and involve users as active participants.

Since Barragán and Goéritz, minimalist architect Peter Zumthor (2006) has also been in the forefront of conceptualizing the link between emotion and architecture. For Zumthor, our emotional connections to a building are related to the user's imaginative appropriation of the space. For him, it is not a question of trying to stir up emotions in persons via buildings, as Barragán and Goéritz suggest, but rather allowing emotions to emerge in the interaction of the building and the user given their singular relationship. It is therefore up to the architect or designer to create a facilitating space that the user can use as they see fit.

Other architects and designers have also brought new understandings in the link between emotion and spaces. The architectural theorist Giuliana Bruno (2011) has notably conceptualized the link between (e)motion and architecture. She claims that architectures are the *vissuto*, the space of lived experiences, traversed by the histories of its inhabitants and transient dwellers, one where the space is both emotional and in motion. Buildings for her are lived (e)motional spaces that are inherently dynamic, crossing the line between dwelling and travelling. From this perspective, paying attention to (e)motional design also means conceiving of user participation, in particular by designing spaces that incorporate the possibility for activity or that involve users directly in their creation.

From the design side, building upon extensive research in neuroscience and philosophy, Donald Arthur Norman (2005) has given a prominent place to emotion in his theory of design. In order to produce a positive emotional impact on the user, he advocates that a designer can tailor his product using three levels of design: visceral, behavioral, and reflective design. These different levels of emotional design are interrelated: while visceral is meant to reflect the immediate reactions we may have toward something, behavioral how we evaluate its use, and reflective design the long-term evaluation we have about the object in our lives. These different levels of emotional design are interconnected and help understand how the user will feel about an object in both the immediate and long-term.

As we can see from this short introduction, designers and architects have consciously sought to establish the link between affects and design. Several important elements can be retained for our discussion. Firstly, Zumthor reminds us that our connections to spaces are involved with imagination, in a complex relationship involving our need for emotional security. This remark will be particularly relevant for the patient's experience in the hospital room, who is dependent on others and may feel a loss of control in a hospital space. This theme is also closely related to Ulrich's point about the need to feel safe in a hospital space (his first level of supportive design). I will also explore how paying attention to light, color, and sound may facilitate greater emotional security and well-being for the patient.

Secondly, following Norman's reflections on emotional design, we can see the way that we feel about objects evolves in different temporalities: it includes our initial attraction or repulsion (visceral design) to an object, but also how they are used and lived (behavioral and reflective design). This subject is important to our discussion, as it shows that even small, so-called insignificant objects such as the hospital bed can take on a great significance for the patient. In addition, our discussion will look at how rooms are designed, lighted, and the use of colors, as it can make all the difference to patient experience.

Thirdly, following Bruno, Barragán, Goéritz, and Zumthor, we can see that an emotional approach involves participation. It means incorporating the possibility of movement within these spaces. As we will see in our discussion, it is possible for spaces and architectures to encourage movement and activity, and not just passivity, particularly when they allow for interaction between designers and users. I will therefore discuss the importance of participatory design to find solutions adapted to the needs of users. As we will see, this discussion is also important to resolve differing perspectives and priorities among users.

7.3 Emotional Well-Being in the Hospital Space

Having introduced the contours of the discussion and the theories that will help to understand the subject, I will now turn to a series of case studies to understand how hospital design can create affective resources and opportunities. In the first section, I will focus on the hospital waiting area and the patient room. In terms of Ulrich's theory of supportive design, we will see how redesign of these spaces may contribute to his first level (greater control) and second level (social support).

7.3.1 The Hospital Waiting Area

The hospital waiting area is the affective heart of the hospital, one that is full of intense worry, anxiety, and fear for oneself and for one's loved ones. It is the gateway to a diagnosis, a treatment, an extended stay outside of home, and one in which one's whole life may be transformed. While the incertitude over what will happen once the person is admitted to the hospital cannot be fully relieved by architecture and design, it can do a lot to either calm or exacerbate it. As busy spaces that are often overcrowded, with noise, bright light, odors, and the presence of other patients and families in crisis, many persons experience hospital waiting areas as stressful (Lamb 2021). They are spaces generally designed for practicability and rarely for comfort or for the affective needs of patients and families. This is in particular the case for waiting areas in the emergency department, but is often the case in other hospital waiting areas as well. To add insult to injury, it has been shown that affects experienced by patients and families in these spaces are often not well detected by hospital staff (Yoon and Sonneveld 2010). There is thus an urgent need to find ways to make them a safer affective space for patients and families.

Hospital management is increasingly aware that such spaces induce stress, which is caused not only by what may happen to them when in care, but also by the prolonged waiting time. Long waiting times have serious effects on care, as they contribute to lack of satisfaction and may even lead to aggression and violence toward healthcare providers (Fryburg 2021). Given these issues, a common strategy by hospital management is to encourage distraction, such as giving access to a television or magazine. While such distractions are thought to make waiting easier, research shows some of them may actually exacerbate stress. This is due in particular to the distraction chosen. Both watching negative events on the news or more neutral programming which may cause boredom have been shown to increase stress in these spaces (Fryburg 2021). While Ulrich's proposal of supportive design showed that positive distractions may facilitate greater well-being for patients, as we can see here, not all distractions are equal. Many hospital waiting rooms—due to their design, including their negative distractions—often provide an *unsafe affective environment*, one which makes patients and families become more stressed, anxious, and worried. This means that even before they have accessed care, their experience is already negative. Let us therefore explore some possibilities in how to redesign these spaces to become affective resources for patients and families.

7.3.2 Biophilic Designs: Inspirations from Nature

One of the most popular design movements at the current time is to bring inspirations from nature to facilitate stress reduction and well-being in indoor spaces. The use of nature in the hospital has long been found to have a positive effect on health and well-being (Ulrich 1984) and in recent years, there has been a growing interest in biophilic design to support these goals. By biophilic design, I refer to those strategies that aim to stimulate the links with nature in both indoor and outdoor spaces. In the context of hospital design, biophilic strategies use nature as a therapeutic facilitator to help alleviate distress, depression, anxiety and even pain (Tekin et al. 2023).

Biophilic design can work by incorporating both direct and indirect experiences of nature. Direct experiences include the incorporation of natural elements and patterns into the design of the built environment, such as designing a plant wall in a hospital waiting room or other ways to bring nature directly indoors, such as views outside the window. Indirect experiences on the other hand integrate natural elements and patterns with elements that mimic or give access to nature, such as photographs or paintings of nature, the use of natural materials and colors, or biomimicry (Jung et al. 2023).

Evidence-based design has shown that both types of biophilic design can contribute to the well-being of its users in the waiting room, although most users will of course prefer a more direct access to nature compared to a photograph. In a study by Beukeboom et al. (2012), researchers showed that users in a waiting room exposed to both real plants (direct biophilic design) as well as a plant poster (indirect biophilic design), reported lower levels of stress. Similarly, a study by Fryburg (2021) investigated how images of plants and nature via various media such as photographs influenced well-being. They demonstrated how indirect biophilic design encouraged greater relaxation and decreased negative emotions in the hospital waiting room. While both of these studies targeted a specific patient population or hospital space, they suggest that both direct and indirect biophilic design can help reduce the stress and anxiety of patients and families in the hospital, serving as affective resources to help the person feel calmer while waiting. They do this in particular by providing positive distraction strategies (Ulrich's third level) which may facilitate more calming effects and relaxation compared to the other more negative distraction strategies such as watching the news.

Furthermore, while the discussion thus far has focused on the hospital waiting area, the benefits of having access to nature goes beyond this stressful space. Nature can be harnessed and integrated into other hospital spaces, with adaptation, as the priorities will depend on the type of patients and their specific needs. For instance, a systematic review by Tekin et al. (2023) showed the importance of identifying and ranking biophilic design parameters across user categories. For outpatients, priorities were access to fresh air, light, thermal comfort and welcoming and relaxing spaces (Ulrich's first and third levels). For inpatients, it was important to feel relaxed and comfortable, have access to light and daylight, but also to feel security and protection (Ulrich's first level). Finally, for staff, the priorities identified were possibilities for privacy and refuge, such as through a staff garden (Ulrich's first level). This finding shows the need to individualize design spaces for the type of user. I will explore this subject in the next section.

7.3.3 *Hybrid Spaces of Privacy and Interaction*

From our discussion thus far, we can see that some design strategies may help create a calmer environment for the individual and in the collective experience of waiting. However, we cannot necessarily expect that all patients will benefit from the same design solutions. Not all patients are the same and may need different affective resources. For instance, while some persons may be calmed by having access to private spaces, others may find it a source of anxiety and will need to be able to communicate with others. However, given the collective experience of a hospital waiting room, is it realistic to tailor design solutions for individual affective needs?

To better understand how such a design may be realistic, let us take Douglas et al.'s (2018) example of a redesign of a public waiting area outside an intensive care unit for families and friends at Auckland City Hospital in New Zealand. The original waiting rooms were located at the center of an overbridge connecting two buildings. They were dark or isolated. They had no windows and uncomfortable chairs and couches. The traffic through these spaces, as well as the lack of light or contact with the outside world, exacerbated the negative emotions (stress, anxiety, and fear) felt by loved ones awaiting news. In their evaluation, the existing space was perceived as "cold, clinical, and impersonal." It failed to provide comfort or privacy to families awaiting—often terrible—news. The authors said that these waiting spaces were, "the cultural and emotional heart of the hospital." Yet, we can see here all of the elements of an unsafe affective space: noise, lack of light, and uncomfortable waiting spaces, all of which made the loved ones' anxiety greater while waiting.

Given these problems, a participatory redesign process was initiated. Its goal was to protect the, "privacy, dignity, and the complex emotional needs of families present…whilst providing an accessible space for staff." Using a participatory approach, the design team first conducted a scoping workshop with users (healthcare professionals and a family member) to identify the problems described above. The second participatory methodology was to exhibit the sketches and illustrations from the scoping workshop in an open space. This step was a way to generate conversation and debate about the best way to redesign this space and to involve new users in the discussion.

In this second step, the researchers noted there were some disagreements among users in their priorities of how to redesign the space, with some interested in providing small family units (cabins), while others prioritized sharing experiences in a collective space. This was largely due to different needs among users, including the need both for a space accessible for staff and a design that facilitated the emotional security of the families waiting. These different needs were to be taken into consideration into the final design solution. The final step of the participatory design process was to create a simple scale model with movable components to allow future users to improve the proposed design. In this step, users were able to move spaces to make them more accessible to staff, to create new areas of privacy or interaction for families, to increase visibility for healthcare staff to find families, etc. While the designers came up with a potential design solution, it was up to the users to make the final design decisions based upon their needs and how they would realistically use the space.

The result of this participatory methodology—and the diverse needs expressed by its users—was the creation of a hybrid space. The space now includes both zones of privacy but also common spaces allowing for movement and interaction. For instance, the design team showed that for some users, it was necessary to create a space in which they could feel a sense of intimacy and security to mitigate the negative emotions experienced by waiting. Others however considered it necessary to enable interaction with others in order to share experiences. This was the case for example of families who felt anxious about their loved ones and were able to engage in social support (Ulrich's second level) and positive distraction (Ulrich's third level) by talking with others in the collective space. The hybrid space allowed both spaces for privacy for quiet contemplation but also for activity and interaction with other users.

As the architect Zumthor has advocated, it remains vital not to impose certain emotions on the user, but rather to provide a facilitating environment. By designing a hybrid space, it was for the user to inhabit it as they wished, desired, or needed, at that particular moment and based upon their needs. This example helps us understand how we may be able to personalize a space to value the user's individual affective needs. As we can see here, the design solutions proposed were not that expensive, but they can make all the difference in whether or not the waiting area was an affective resource or a constraint for its users. Furthermore, as we have seen from the methodology proposed, it was possible to articulate and adapt to multiple needs (functional from the healthcare provider perspective and affective from the family perspective) by involving them in the design process.

7.4 The Hospital Room

Having explored two examples of how to redesign the hospital waiting area to promote well-being, let us now turn to the care experiences of inpatients who stay in a hospital room. During the period of hospitalization (long or short), a patient is confined most of the time to a hospital room, and in particular to their hospital bed. For the patient who is bed-bound in a hospital, their room is temporarily their whole world. Within these spaces of limited movement, the vastness of the hospital is almost unimaginable. The corridors, closed and open rooms, numerous reception and waiting areas that are humming twenty-four hours a day are largely inaccessible to the person who only sees fragments of this world in quick journeys from their room to the treatment room, corridors, and waiting areas. These spaces are immense and to a certain extent, *unimaginable*, for the person when their mobility and movement is restricted. Because of this, the hospital room may take on increased significance, as it represents the safe space from which they experience the hospital. This space can either be threatening if it adds to the fears and anxieties of the patient, or it can help facilitate feelings of calmness and reflection if the patient can experience it as secure (Ulrich's first level of supportive design).

Furthermore, in thinking about how the hospital room may serve as an affective resource, there is no need to idealize this space, which represents an unstable space for intimacy and privacy for the patient due to its functional needs. First of all, it is accessible at all times to healthcare staff. It is also in many cases a shared space with other patients. Therefore, these spaces will never be close to a home environment. However, as NoorMohammadi (2015) has shown, when we experience a place that does not relate to our values of being (for instance if it is too cold, too modern, without light or color), we may not feel safe or secure, and this may exacerbate fear and anxiety. Let us explore these issues by looking at various furniture in the hospital room and how it may serve as affective resources or constraints for patients.

First of all, probably the most important object in the hospital space is the hospital bed. It is the place where patients will spend the most time, that they feel (un)comfortable or un(safe) in depending on their health condition and the bed's design. Daniel Norman proposed that we should consider the different levels (visceral, behavioral, reflective) of emotional design in such objects. From the patient perspective, the bed is the place that can serve a functional use for the user by being both accessible and useful, but also one that can help the person feel secure or insecure in the hospital room. At first the hospital bed—given its unique design for functional and care needs—may be imposing for its future user and even create fear (a visceral reaction to the object). While designers have done their upmost to make it attractive, it is still a heavy and imposing object that at first sight gives none of the comfort of a traditional bed. However, when the patient grows used to it during their hospital stay, they will either come to see it a safe space (as it both comfortable but also promotes a sense of safety and security), or they will conceive of it as an unsafe space, particularly when they have had a fall or another problem due to the bed design.

To better understand how a bed may become an affective resource for the patient, we will take an example by Wiggermann et al. (2019), who designed a bed for use in medical-surgical units, intensive care units, and long-term acute care units. Given the complex needs of these users, the bed design had to be useful and practical for healthcare staff (nurses, nurse assistants, physical therapists, housekeeping staff, etc.)—but also a resource for the patient who had to feel comfortable and secure in it. In addition, because of the complex care needs of patients in these units, and in particular their dependance, their bed is practically their whole world. They have to engage in diverse activities in it throughout the day, whether it be eating, sleeping, cleaning, positioning for therapy, being weighed, being transported in the hospital, or even performing emerging procedures such as CPR (Wiggermann et al. 2019).

This bed was first of all designed to take into consideration practical factors in the transport and care for the patient from the bedside, which made it an accessible and easy-to-use object from the healthcare provider's perspective. From the patient perspective, however, other needs also needed to be taken into consideration, and in particular how to move from their bed to other spaces in the room, such as the bathroom or chair. Patients interviewed were in particular concerned with fall prevention. Their study showed that while these patients typically try to call the nurse for help to get out of bed, the nurse call was often disconnected or unfunctional. In many bed designs in a hospital, this is a common problem. Because of untransparent design, patients often had no way to know if their call was connected or answered. When they did not receive a response, some patients tried to get out of bed independently, which increased their risk of a fall.

Because of these concerns, Wiggerman et al. (2019) proposed several design solutions to increase patient perceptions of safety and control. The first strategy was to give patients greater control over their contact with the nurse. They designed the nurse call button's functionality to be clearly visible to the patient: when the button was illuminated, they could be confident that their call was placed and that the nurse would respond. This encouraged them to wait for the nurse before getting up independently. The second way the bed design was tailored to their needs was to create a lighted side-rail grip that provided an easy visual reference for getting in and out of bed. The grip remained illuminated while the patient moved from the bathroom or to their bedside chair, giving them an easy reference to find the grip again upon return. This was useful when the patient got up during the night and had to regain their bed again afterwards.

Thanks to these two features, the bed was perceived of as safer to its users and gave them a greater sense of control of their mobility decisions (Ulrich's first level). I have chosen to discuss these features as they are those prioritized from the patient perspective. In addition, they show some of the ways that the design of such an object can influence patient experience, and in particular how it may increase their autonomy. By promoting a greater sense of control (through the nurse call button) and practical features to promote patient safety (the lighted hand rail), they helped patients to feel less anxious when they had to be mobile in a period of intense vulnerability and risk for them.

This example shows that functional objects such as the bed are of great importance to the patient in their hospital experience, and that with design, they either become safe or unsafe affective resources for the patient. While to our knowledge there is no evidence-based design studies that look at the other objects in the hospital room, let us take a moment to consider them with this perspective. For instance, the closet and/or the drawers are mostly seen in a purely functional capacity as the place to keep personal affairs and to reduce the possibility of theft. However, when thinking about questions of control (Ulrich's first level), it is possible to understand the symbolic importance of these objects to the patient. First of all, they are the means for privacy in times of emotional upheaval, especially if they are secured (i.e. inaccessible to other hospital users), which may help patients to feel a greater sense of control. While these kinds of "inconsequential" objects or small spaces may not seem important in the context of increasingly expensive and technical healthcare acts, they may take on an importance to the person and help them inhabit these spaces, even if in a short-term temporality. Ulrich's levels of supportive design also brings up the question of safety and how such objects can be better designed for the user. Seen from the example of the hospital bed, an important feature was the possibility for safe mobility. We can understand that these objects may become fearful to the patient if they are inaccessible or not easily reached, particularly when they may cause a fall due to their inaccessibility or placement in a difficult to reach part of the room. While we can expect that the family member or healthcare provider may provide access to such spaces, making them easily accessible (and safe) for the patient may increase their sense of autonomy and control.

These are some of the ways that object design serves as an affective resource for the patient. In addition, as we know from our conversation on biophilic design, the patient's room may provide a sense of comfort and privacy when patients can engage in positive distraction through nature (Ulrich's third level of supportive design). Indeed, Ulrich's (1984) most famous work showed how simply having a view through a window could influence recovery from surgery. In this emblematic study, he showed that those patients who had rooms with windows looking out on a natural scene had shorter postoperative hospital stays, but also received fewer negative evaluations from nurses and took fewer strong painkillers than in those with windows that faced a brick wall. While we can understand that a room with a view may not always be feasible, our discussion on biophilic design showed that even indirect biophilic design, such as images of nature, may also help patients to benefit.

From this discussion, we can see that the design of objects in the hospital room, and even so-called inconsequential objects such as the hospital bed or chest of drawers—may make a difference to the patient in how they experience care. While for the most part their purpose is seen purely in a functional capacity, they can also be designed to become affective resources for the patient. They can do this by providing a way for the user to feel protected, more in control, and safe. Having explored how the design of objects may provide affective resources for patients and families, let us now turn to how sound, light, and color may influence patient experience. As I have advocated, making these spaces more hospitable needs a holistic approach, one in which we consider the objects in the room, access to nature, but also how sound, light, and color may influence patient experience.

7.5 Emotional Design: Paying Attention to Sound, Light, and Color

In recent years, evidence-based design has been helping make hospitals safer for patients, families, and staff by taking into account factors such as lighting, contact with nature, and noise. This has helped to reduce medical errors, hospital-based infections, and improve treatment outcomes (Ulrich et al. 2008). While these are important factors in hospital design and architecture, the influence of sound, light, and color on the patient's overall affective experience is not to be minimized. In this section, I will in particular focus on the hospital room, as this is where the patient will likely be most affected by sound, light, and colors; however, I will also look at other spaces, in particular hospital corridors, to better understand how user experience can be improved.

7.5.1 The Auditory World

Research in emotional design has shown that sound can be emotionally satisfying and informationally rich (Norman, 2005). However, it can also be a source of emotional stress. In the hospital, sounds that are conceived of as disturbing or annoying can dominant patient experience (Oleksy and Schlesinger 2019; Xyrichis et al. 2018). If there is limited auditory privacy in the hospital room, for instance due to another patient, it may aggravate the affects—whether it be fear, anxiety, or frustration—that the patient is experiencing. While hospitals may provide short-term functional solutions such as foldable screens or curtains to facilitate visual privacy in a shared room, auditory privacy in these spaces has been largely understudied and undervalued. Similarly, outside sounds from the hallway—such as healthcare providers or families talking—or other abrasive sounds such as hospital monitors or other user's telephones, will inevitably affect patient experience. A study by Akansel and Kaymakçi (2008) showed that the sounds in an intensive care unit most disturbing to patients were the noises created by other patients, those admitted from the emergency room to the unit, monitor alarms, and conversations among healthcare providers. Patients who had a bed located closer to the nurses' station had the highest level of noise disturbance for these reasons. Furthermore, the sounds that the patient hears while being transported from their hospital room to a treatment room can also increase their anxiety. As we can see from this discussion, abrasive sounds in the hospital environment may put the patient in a situation of vulnerability and increase their fear and stress.

However, while these findings suggest that noises in general should be eliminated, creating silence alone may not necessarily create a positive affective environment for the person. At the current time, studies in evidence-based design have a number of contradictory conclusions that show both the positive and negative effects of sound in the hospital. However, Mackrill et al. (2013)'s research can help us to unblock this difficulty. These authors propose that it is important to study the person's overall emotional response to the *soundscape* to better understand the user's affective experience, and in particular whether or not it will be positive or negative. By soundscape, he is referring to the combination of sounds that comes from immersion in an environment. The soundscape in the hospital can come from natural sounds (birds chirping outside) but also environmental sounds created by humans and their technologies, such as telephones or talking. Research on soundscape is interested in the person's perception of sounds and how they are affected by them. While the person may be affected by individual sounds (birds chirping), the overall soundscape (the interaction of sounds amongst each other) was the most important factor in user experience. While it is unlikely that natural sounds (particularly if they are outside) can be controlled, environmental sounds can be at least partly regulated through hospital design.

Mackrill et al. (2013) show how the soundscape can be healing to the user, and they advocate that a silent environment is not necessarily the solution for all. Their study showed that for some users, the "gentle hum" of people doing things can reduce isolation, and for some, too much quiet may cause anxiety. This lends support to the idea that not all sounds in the hospital are alike. There is increasing evidence that some sounds may encourage feelings of calm and well-being in the hospital space. While intrusive sounds such as those coming from telephones represent the kind of negative distractions that may increase anxiety or frustration, other distractions may be perceived as more positive to the user (Ulrich's third level). An example of a positive distraction is music. Studies show that the use of music in the hospital (when adapted to the needs and tastes of its users) can both help reduce anxiety and stress, but also make patients feel more comfortable and secure. In addition, when patients are able to choose which music to prioritize, they may feel greater control of their environment (Ulrich's first level). Furthermore, some sounds or music are likely to be more healing than others. For instance, research has shown that listening to soothing music such as sounds of nature (winds, birds, ocean sounds) in particular may reduce stress, blood pressure, and post-operative trauma compared to silence (Iyendo 2016).

These findings lend support to the idea that sound can be purposely harnessed to improve hospital experience and provide affective resources for the patient. Better understanding the overall soundscape in these spaces from a person-centered perspective will therefore help hospital designers and administrators decide what sounds should be minimized or eliminated. As Xyrichis et al. (2018) advocate, "patients and relatives are rarely involved in this research; they are seen as passive recipients of the hospital soundscape rather than active participants in its creation…future solutions should be based on a careful assessment of each hospital environment, and designed with input from all stakeholders, particularly patients and families."

Finally, it is not to be forgotten that adapting the overall soundscape will also affect the healthcare professional. As we saw in the discussion on biophilic design, a priority for staff was the possibility for privacy and refuge (Ulrich's first level). While some hospitals have taken steps to ensure separate spaces for staff, auditory privacy is a little studied issue, with important effects on healthcare well-being and performance. It has been shown that noise pollution leads to stress and burn-out among healthcare workers (Walker and Karl 2019; Topf 2000). For instance, a study by Deniz and Sarac (2023) indicated that 95% of their respondents believe hospital noise affected their performance, with direct or indirect adverse health outcomes for both patients and healthcare staff. Hospitals are noisy environments for healthcare workers, with sounds generated by medical devices, pagers, other staff, patients and their relatives. Studies have found that noise levels in hospital exceeded those recommended by the World Health Organization, with recommended noise levels not to exceed 30 dB. In the study by Deniz and Sarac (2023) for instance, respondents recorded noise levels between 37.4 dB and 73.7 dB(A), which is well above the WHO's recommendations. As we can see here, noise levels in the hospital not only affect patient experience, but healthcare provider experiences and performance as well.

As our discussion on biophilic design showed, soundproofing spaces—as well as providing respite where healthcare providers can have privacy, including auditory privacy—is a way to promote healthcare provider well-being (Gregory 2021). One of the ways to do this is to provide outside nature spaces (such as a garden on the roof) specifically devoted to hospital staff, which will provide a refuge to healthcare providers and serve as affective resources on multiple levels—from nature, from quieter soundscapes. In a similar way, studying the impact of calming soundscapes in hospital services or corridors—such as the use of nature sounds—may also help to improve the auditory world for the healthcare provider. However, it will not be a question of merely imposing one sound over another, for instance birds chirping to mask the sound of telephones. This kind of solution may not only make the overall soundscape louder, but it may also affect patient safety, particularly when hospital sounds alert to a problem. Solutions will therefore always need to be considered in line with and not in opposition to these more functional needs.

7.5.2 Color

In addition to sound, color may also impact our affective experience in hospitals. While there has historically been a tendency to use white colors in hospital spaces to encourage associations with cleanliness, hospitals are increasingly using other colors to soothe hospital users. Among designers and architects, however, there is no consensus that any one color should be prioritized. Cultural and individual differences in how a particular color is perceived makes it difficult to find a universal solution to this issue. However, increasing data is making the case that in hospitals: (1) the use of colors is better than the use of white; (2) some colors or color combinations may be more effective than others in promoting patient well-being.

In terms of the first issue, a study by Eminovic et al. (2022) supports the hypothesis that color vs. white can promote greater well-being in the hospital space. Their intervention research, consisting of a prospective randomized controlled study of 80 patients facing total hip or knee arthroplasty, compared quality of life and mood of those who stayed in a white room (control) group vs. those rooms that were designed with six light color codes thought to have calming effects. Their study demonstrated that while preoperative anxiety was high for both groups awaiting the surgical procedure, there was greater perceived quality of life for the intervention group than the control group during postoperative care. As advocated by the authors of the study, the use of colors in hospital rooms can therefore be carefully planned and studied in order to promote greater patient well-being, particularly in some temporalities of patient care (recovery). Their study also shows the use of colors is an effective and low-cost intervention that can change how the patient and family perceives the hospital space.

In terms of the second issue, which looks at whether one color should be prioritized over another, it is unlikely there will be a consensus on whether one color should be used over another. However, an exception to this issue is likely the use of green due to its association with nature. I have already explored the potential of using nature to help provide an affective resource in these spaces. Designs that encourage the use of green—particularly ones that make the link between plants or other green landscapes through drawing, photographs, or other types of media—may create this association to nature that helps to calm. It is for this reason that the color green has been increasingly used in the hospital space.

Another potential for the use of colors is to promote feelings of safety and control. Colors are increasingly being used in the hospital to help the user navigate the space (Olguntürk et al. 2023). This second issue is of importance to all patients and families, who often have trouble to navigate a large hospital space; however, it will be of particular importance for elderly and visually impaired people, as it may help them to find clear visual cues. I can hypothesize here, following Olguntürk et al. (2023), that by making the hospital easier to navigate, it may help to reduce the stress of these hospital users and help them have greater control of their environment (Ulrich's first level).

This brings us to the issue of tailoring colors for individual patient group needs. For instance, while the elderly may benefit from visual clues to navigate the hospital space, young children have special needs related to their early childhood development. In particular, it has been found that young children make associations with color more than their forms. In practice, this means designing a colorful playground area can help reduce negative emotions experienced by children in the hospital space due to fear and boredom (Olguntürk et al. 2023). In this way, it can provide the type of positive distraction (Ulrich's third level) adapted to the child's needs.

These are at least three ways in which colors may provide affective resources for patients. As we can see from this discussion, it is important to tailor the use of color to the needs of individual patients—or at least groups of patients—and make these choices part of the overall design to create a more soothing environment for users.

7.5.3 Light

In addition to sound and colors, light influences the affective experience of the patient. The effects of light on the user have been studied since the advent of environmental design. Research has notably emphasized how natural light can create an emotionally smoothing environment and facilitate the circadian rhythm. I have already discussed Ulrich's seminal article (1984) that showed how having a window view of a natural setting had restorative influences on surgical patients and increased their satisfaction. Likewise longitudinal studies have shown that the patient's length of stay is shorter for those nearer to a window compared to a door in multi-occupancy rooms (Park et al. 2018).

Raanaas et al. (2015) have also demonstrated how access to natural light and views of nature can provide more emotional security in a hospital room. Their study of patient behavior in a residential rehabilitation center demonstrated that those with access to natural light and a view of nature through a window had improved self-reported health. This was because they had a tendency to use their room as a space in order to cope emotionally with the demands of their rehabilitation program. Given that these patients were taxed by the limitations of their health and the tiring effects of the program, seeing their rooms as affective safe spaces appeared to have a positive effect on their well-being and recovery.

This is not to suggest however that light is always a good thing. Too much light can aggravate negative emotions, as a hospital that is always bright does not necessarily allow comfort for the patient in need of rest and reflection. Many hospitals still use fluorescent tube lights in patient rooms, which has been shown to interfere with sleep, rest, and recovery. In a similar way, the use of lights during nighttime in the hospital may also interfere with the patient's sleep. Healthcare providers often use flashlights and mobile telephone lights at nighttime in order to avoid disturbing patients, not realizing that these light sources are particularly disruptive (Albala et al. 2019). Soliciting user experience, for instance by using self-report questionnaires (Lindahl et al. 2021), may help understand how to tailor light and color to the individual's emotional needs.

In these examples, I have emphasized the need to reflect on how light and color can be tailored to the individual person's affective needs. Before concluding this section, let me take a moment to better understand how smart environment research may help us move in this direction. This field is rapidly developing methods to adapt spaces to induce positive emotional states in their users through technology. Fernández-Caballero et al. (2016) for instance have shown that it is possible for a sensor to detect an emotional state in persons by analyzing their physiological signals, facial expressions, and behavior. The sensor will then adapt to the person's emotional state by providing emotion regulation through music and color/light in a given space. It can be questioned whether emotions can be detected accurately via facial expressions and whether different persons will necessarily be affected in the same way by different colors, lights, or sounds. However, smart environment research is expected in the future to be able to effectively tailor their solutions to individual needs. From this perspective, they are a promising affective opportunity to enable users to have design solutions tailored to their needs.

However, the use of smart technologies also raises other issues from the patient perspective, in particular how the use of sensors—and other forms of technologies which may be viewed as invasive—may affect the user's overall affective experience. While the current discussion in this area mostly focuses on ethical questions, such as whether users are willing to accept the gathering of personal data, such as images, temperatures, and facial expressions, and whether such data may be biased as based upon a select data set, the debate can go farther by investigating how technologies may affect patient perceptions of care. It can notably help to better understand whether these technologies facilitate or exasperate the user's emotional security in these spaces.

In this section, I have explored how sounds, light, and color can affect user experiences, and in particular in the patient room. I have also explored the possibilities and limits of the use of technology to tailor for the individual's affective needs. I will now explore how hospital spaces may be adapted for greater activity and interaction.

7.6 Rethinking Hospital Spaces for Activity and Interaction

In this final section, I will explore some ways to find a balance between rest and activity in hospital spaces following Bruno's ideas on *vissuto*. To start with, hospitals are dynamic spaces. Encounters between hospital staff and patients form the lived experience and the (e)motions lived in these encounters. Healthcare providers and their support staff are constantly moving between patient rooms, staff rooms, and places of technical acts, discussing with patients and being discussed by them. Patients are shuffled by healthcare staff between their rooms to spaces where technical analysis and treatment awaits them. Families, friends, and other visitors are those transient dwellers who come to visit and wait with patients and often interact with healthcare staff. According to Bruno's reflections on *vissuto*, an attention to (e)motional design will therefore need to incorporate both the possibility of rest but also a space for activity, interaction, and movement for its users in ways that do not disturb the safe spaces created through other methods. Let us therefore first of all discuss how hospital spaces may be redesigned for interaction before moving on to more participatory methodologies to give users a greater say in design decisions.

7.6.1 Redesigning Hospital Spaces for Interaction

Patients in a hospital are often confined to a bed and have limited capacity for movement. While design can help create a safe space for the patient, it can also become an oppressing space for them due a lack of human connection and lead to sadness, anxiety, and feelings of isolation. While there are often common spaces in long-term care facilities such as in retirement homes or reeducation centers, there is very little space designed for patient activity in the hospital service. As Feenstra (2021) has shown, introducing new rhythms of activation, relaxation, and sleep can help change the hospital experience from almost completely passive to one with periods of activity, and by these means help patients toward recovery. Providing common spaces may also benefit the patient by helping them to inhabit more spaces in the hospital and to interact with other visitors or patients. Such interactions may help to reduce their feelings of sadness, anxiety, and isolation, as well as help them have more perceived control of their environment.

7.6.2 Participatory Design Methodologies

Participatory methodologies may also help users move from passivity to activity. If they are designed without the active participation of its current and future users, we risk to make them uninhabitable for patients, families, and healthcare providers. Valuing affects in hospital spaces therefore also means valuing the active participation of its users, be it from the design of "inconsequential" objects such as hospital closets to the redesign of waiting areas, hospital rooms, and gardens. To do in this in a practical way, Oel et al. (2021) has proposed that we take into account the differing ways of communicating in design decisions between hospital users (mainly verbal or written expression) and architects and designers (mainly visual expression). Likewise, Devlin et al. (2016) have shown the importance of qualitative approaches to better understand how hospital design, in particular in the hospital room, affects patients' well-being and satisfaction, while Douglas et al. (2018) give several methods of how to go about participatory design. I have given some suggestions throughout this chapter on ways to test design decisions with users, in particular through qualitative methods, which allows users to describe and elaborate their ideas in their own words. However, whatever the method ultimately chosen, taking a person-centered perspective is necessary to ensure that architecture and design decisions fulfill the user's affective needs. To do so, hospital administrators, designers, and architects will need to work *with* patient groups, patients, families, and healthcare workers to capture how these spaces are lived emotionally. Table 7.2 summarizes some participatory methods to engage users in design or redesign of hospital spaces.

As seen throughout this chapter, it is be expected that different users may have different needs (functional, affective) that may influence how they live a space. Ulrich (1991) highlights that different groups—patients, visitors, or staff—sometimes have conflictual needs with respect to his three categories of control, social support, or positive distraction. He gives the example of the receptionist in the waiting area who may wish to control the programs on a television and the patient who may experience

Table 7.2 Participatory methodologies and person-centered design

Author	Methodology	Purpose	Who can be involved?
Devlin et al. (2016)	Qualitative interviews/ focus groups or observations of the existing space and how it is used	Better understand user needs	Healthcare providers, families, and patients
Douglas et al. (2018)	Scoping workshop	Define needs, priorities, and problems	Healthcare providers, families, patients, management, and other users passing through the space
	Exhibition of design sketches and illustrations	Stimulate conversation on design propositions	
	Small-scale adjustable model that can be manipulated by users	Give an active role for future users in improving initial design solutions	
Lawson (2010)	IDEAs (Inspiring Design Excellence and Achievements) web-based tool	Link research evidence to the challenges, considerations, and needs of users	Clients, users, architects

greater stress due to a loss of control. Similarly, he discusses how bright colors in hospital corridors and rooms may be more agreeable to healthcare providers, whilst increasing stress for certain users who need to use their rooms as spaces of privacy and reflection. As seen throughout this chapter, reconciling different affective needs and priorities is no easy task. However, as seen in Douglas et al. (2018)'s example of a hybrid space, compromise is possible, particularly when participatory methodologies are used to discuss, negotiate, and co-create hospital spaces and architectures.

7.7 Conclusion

This chapter has advocated taking seriously how spaces and architectures are lived emotionally by its users, and the ways in which they can become affective resources for them. Input from emotional designers and architects, as well as research in environmental design, have helped conceptualize ways to provide them, even in a busy hospital space. Some of these proposals may involve an investment in redesigning hospital spaces and architectures to make these spaces more habitable, such as a hospital waiting area or redesigning the building to allow greater access or views on nature. However, many of these solutions will be simple and inexpensive, such as adapting the patient's soundscape to make these spaces more emotionally secure. However, in all cases, design decisions will need to include the active involvement of patients, families, and healthcare providers to resolve competing priorities and to find ways to design the hospital so that they are both practicable and respond to the affective needs of their users.

References

Akansel, Neriman, and Şenay Kaymakçi. 2008. "Effects of Intensive Care Unit Noise on Patients: A Study on Coronary Artery Bypass Graft Surgery Patients." *Journal of Clinical Nursing* 17 (12): 1581–1590. https://doi.org/10.1111/j.1365-2702.2007.02144.x

Albala, Lorenzo, Timothy Bober, Graham Hale, Benjamin Warfield, Micaela Langille Collins, Zak Merritt, Eric Steimetz, Shmuel Nadler, Yair Lev, and John Hanifin. 2019. "Effect on Nurse and Patient Experience: Overnight Use of Blue-Depleted Illumination." *BMJ Open Quality* 8 (3): e000692. https://doi.org/10.1136/bmjoq-2019-000692

Beukeboom, Camiel J., Dion Langeveld, and Karin Tanja-Dijkstra. 2012. "Stress-Reducing Effects of Real and Artificial Nature in a Hospital Waiting Room." *The Journal of Alternative and Complementary Medicine* 18 (4): 329–333. https://doi.org/10.1089/acm.2011.0488

Bogaert, Brenda. 2021. "Moving Toward Person-Centered Care: Valuing Emotions in Hospital Design and Architecture." *HERD: Health Environments Research & Design Journal*, December, 193758672110621. https://doi.org/10.1177/19375867211062101

Bruno, Giuliana. 2011. *Atlas of Emotion: Journeys in Art, Architecture, and Film*. Paperback ed., Reprint. New York, NY: Verso.

Dannenberg, Andrew L., and Heather Burpee. 2018. "Architecture for Health Is Not Just for Healthcare Architects." *HERD: Health Environments Research & Design Journal* 11 (2): 8–12. https://doi.org/10.1177/1937586718772955

Deniz, Huseyin, and Elif Sarac. 2023. "Noise Exposure in Healthcare Workers." *Medicine Science | International Medical Journal* 12 (4): 1232. https://doi.org/10.5455/medscience.2023.08.127
Devlin, Ann Sloan, Cláudia Campos Andrade, and Diana Carvalho. 2016. "Qualities of Inpatient Hospital Rooms: Patients' Perspectives." *HERD: Health Environments Research & Design Journal* 9 (3): 190–211. https://doi.org/10.1177/1937586715607052
Dijkstra, Karin, Marcel Pieterse, and Ad Pruyn. 2006. "Physical Environmental Stimuli That Turn Healthcare Facilities into Healing Environments through Psychologically Mediated Effects: Systematic Review." *Journal of Advanced Nursing* 56 (2): 166–81. https://doi.org/10.1111/j.1365-2648.2006.03990.x
Douglas, Reid, Stephen Reay, Josh Munn, and Nick Hayes. 2018. "Prototyping an Emotionally Responsive Hospital Environment." *Design for Health* 2 (1): 89–106. https://doi.org/10.1080/24735132.2017.1412689
DuBose, Jennifer, Lorissa MacAllister, Khatereh Hadi, and Bonnie Sakallaris. 2018. "Exploring the Concept of Healing Spaces." *HERD: Health Environments Research & Design Journal* 11 (1): 43–56. https://doi.org/10.1177/1937586716680567
Eminovic, Sandra, Gabor Vincze, Andrea Fink, Stefan F. Fischerauer, Patrick Sadoghi, Andreas Leithner, Lars-Peter Kamolz, Karlheinz Tscheliessnigg, and Gerwin A. Bernhardt. 2022. "Positive Effect of Colors and Art in Patient Rooms on Patient Recovery after Total Hip or Knee Arthroplasty: A Randomized Controlled Trial." *Wiener Klinische Wochenschrift* 134 (5–6): 221–226. https://doi.org/10.1007/s00508-021-01936-6
Feenstra, Femke. 2021. "Reactivating Hospital, Active Patients Through Special Design." *HERD: Health Environments Research & Design Journal* 14 (1): 87–105. https://doi.org/10.1177/1937586720966757
Fernández-Caballero, Antonio, Arturo Martínez-Rodrigo, José Manuel Pastor, José Carlos Castillo, Elena Lozano-Monasor, María T. López, Roberto Zangróniz, José Miguel Latorre, and Alicia Fernández-Sotos. 2016. "Smart Environment Architecture for Emotion Detection and Regulation." *Journal of Biomedical Informatics* 64 (December): 55–73. https://doi.org/10.1016/j.jbi.2016.09.015
Fryburg, David A. 2021. "What's Playing in Your Waiting Room? Patient and Provider Stress and the Impact of Waiting Room Media." *Journal of Patient Experience* 8 (January): 237437352110498. https://doi.org/10.1177/23743735211049880
Gallegos, Francisco. 2021. "Affective Injustice and Fundamental Affective Goods." *Journal of Social Philosophy* n/a (n/a). https://doi.org/10.1111/josp.12428
Goéritz, Mathias. 1953. "Emotional Architecture Manifesto." https://designmanifestos.org/mathias-goeritz-emotional-architecture-manifesto/
Gregory, Debbie. 2021. "Code Lavender: Designing Healthcare Spaces to Enhance Caregiver Wellness." *HERD: Health Environments Research & Design Journal* 14 (2): 13–15. https://doi.org/10.1177/1937586721993785
Iyendo, Timothy Onosahwo. 2016. "Exploring the Effect of Sound and Music on Health in Hospital Settings: A Narrative Review." *International Journal of Nursing Studies* 63 (November): 82–100. https://doi.org/10.1016/j.ijnurstu.2016.08.008
Jung, Dawoon, Da In Kim, and Nayeon Kim. 2023. "Bringing Nature into Hospital Architecture: Machine Learning-Based EEG Analysis of the Biophilia Effect in Virtual Reality." *Journal of Environmental Psychology* 89 (August): 102033. https://doi.org/10.1016/j.jenvp.2023.102033
Lamb, Matthew D. 2021. "Health Inequity by Design: Waiting Rooms and Patient Stress." *Frontiers in Communication* 6 (September): 667381. https://doi.org/10.3389/fcomm.2021.667381
Lawson, Bryan. 2010. "Healing Architecture." *Arts & Health* 2 (2): 95–108. https://doi.org/10.1080/17533010903488517
Lindahl, Jeanette, Hans Thulesius, Mikael Rask, Helle Wijk, David Edvardsson, and Carina Elmqvist. 2021. "Assessing the Supportiveness of Healthcare Environments' Light and Color: Development and Validation of the Light and Color Questionnaire (LCQ)." *HERD: Health Environments Research & Design Journal* 14 (2): 130–144. https://doi.org/10.1177/1937586720975209

MacAllister, Lorissa, Craig Zimring, and Erica Ryherd. 2019. "Exploring the Relationships Between Patient Room Layout and Patient Satisfaction." *HERD: Health Environments Research & Design Journal* 12 (1): 91–107. https://doi.org/10.1177/1937586718782163

Mackrill, J., R. Cain, and P. Jennings. 2013. "Experiencing the Hospital Ward Soundscape: Towards a Model." *Journal of Environmental Psychology* 36 (December): 1–8. https://doi.org/10.1016/j.jenvp.2013.06.004

NoorMohammadi, Susan. 2015. "The Role of Poetic Image in Gaston Bachelard's Contribution to Architecture: The Enquiry into an Educational Approach in Architecture." *Environmental Philosophy* 12 (1): 67–86.

Norman, Donald A. 2005. *Emotional Design: Why We Love (or Hate) Everyday Things*. New York: Basic Books.

Oel, Clarine J. van, Meloek Mlihi, and Arno Freeke. 2021. "Design Models for Single Patient Rooms Tested for Patient Preferences." *HERD: Health Environments Research & Design Journal* 14 (1): 31–46. https://doi.org/10.1177/1937586720937995

Oleksy, Anthony J., and Joseph J. Schlesinger. 2019. "What's All That Noise-Improving the Hospital Soundscape." *Journal of Clinical Monitoring and Computing* 33 (4): 557–562. https://doi.org/10.1007/s10877-018-0215-3

Olguntürk, Nilgün, Rengin Aslanoğlu, and Begüm Ulusoy. 2023. "Color in Hospitals." In *Encyclopedia of Color Science and Technology*, edited by Renzo Shamey, 400–403. Cham: Springer International Publishing. https://doi.org/10.1007/978-3-030-89862-5_449

Park, Man Young, Choul-Gyun Chai, Hae-Kyung Lee, Hani Moon, and Jai Sung Noh. 2018. "The Effects of Natural Daylight on Length of Hospital Stay." *Environmental Health Insights* 12 (January): 117863021881281. https://doi.org/10.1177/1178630218812817

Pellerin, Denis, and Marie Coirié. 2017. "Design et hospitalité : quand le lieu donne leur valeur aux soins de santé." *Sciences du Design* 6 (2): 40–53.

Pelletier, Louise. 2008. "Modeling the Void: Mathias Goeritz and the Architecture of Emotions." *Journal of Architectural Education* 62 (2): 6–13. https://doi.org/10.1111/j.1531-314X.2008.00234.x

Pierron, Jean-Philippe. 2019. "Pour une philosophie de l'attention spatiale. Architecture des soins et soin de l'architecture." In *Cahiers thématiques: architecture et paysage, conception, territoire, histoire, matérialité*.

Pierron, Jean-Philippe, and Didier Vinot. 2020. "The Meaning of Value in 'Person-Centred' Approaches to Healthcare." *European Journal for Person Centered Healthcare* 8 (2): 193–200. https://doi.org/10.5750/ejpch.v8i2.1842

Raanaas, Ruth Kjærsti, Grete Patil, and Grete Alve. 2015. "Patients' Recovery Experiences of Indoor Plants and Viewsof Nature in a Rehabilitation Center." Edited by Randi W. Aas, Ruth K. Raanaas, and Lynn Shaw. *Work* 53 (1): 45–55. https://doi.org/10.3233/WOR-152214

Tekin, Bekir Huseyin, Rhiannon Corcoran, and Rosa Urbano Gutiérrez. 2023. "A Systematic Review and Conceptual Framework of Biophilic Design Parameters in Clinical Environments." *HERD: Health Environments Research & Design Journal* 16 (1): 233–250. https://doi.org/10.1177/19375867221118675

Topf, Margaret. 2000. "Hospital Noise Pollution: An Environmental Stress Model to Guide Research and Clinical Interventions." *Journal of Advanced Nursing* 31 (3): 520–528. https://doi.org/10.1046/j.1365-2648.2000.01307.x

Ulrich, R. 1984. "View through a Window May Influence Recovery from Surgery." *Science* 224 (4647): 420–421. https://doi.org/10.1126/science.6143402

Ulrich, R. S. 1991. "Effects of Interior Design on Wellness: Theory and Recent Scientific Research." *Journal of Health Care Interior Design: Proceedings from the ... Symposium on Health Care Interior Design. Symposium on Health Care Interior Design* 3: 97–109.

Ulrich, Roger S., Craig Zimring, Xuemei Zhu, Jennifer DuBose, Hyun-Bo Seo, Young-Seon Choi, Xiaobo Quan, and Anjali Joseph. 2008. "A Review of the Research Literature on Evidence-Based Healthcare Design." *HERD: Health Environments Research & Design Journal* 1 (3): 61–125. https://doi.org/10.1177/193758670800100306

Walker, Linda, and Cherry A. Karl. 2019. "The Hospital (Not So) Quiet Zone: Creating an Environment for Patient Satisfaction Through Noise Reduction Strategies." *HERD: Health Environments Research & Design Journal* 12 (4): 197–202. https://doi.org/10.1177/1937586719839229
Wiggermann, Neal, Kelli Rempel, Robert Mark Zerhusen, Travis Pelo, and Nick Mann. 2019. "Human-Centered Design Process for a Hospital Bed: Promoting Patient Safety and Ease of Use." *Ergonomics in Design: The Quarterly of Human Factors Applications* 27 (2): 4–12. https://doi.org/10.1177/1064804618805570
Xyrichis, Andreas, John Wynne, Jamie Mackrill, Anne Marie Rafferty, and Angus Carlyle. 2018. "Noise Pollution in Hospitals." *BMJ* 363 (November): k4808. https://doi.org/10.1136/bmj.k4808
Yoon, JungKyoon, and Marieke Sonneveld. 2010. "Anxiety of Patients in the Waiting Room of the Emergency Department." In *Proceedings of the Fourth International Conference on Tangible, Embedded, and Embodied Interaction*, 279–286. Cambridge Massachusetts USA: ACM. https://doi.org/10.1145/1709886.1709946
Zumthor, Peter, Maureen Oberli-Turner, and Catherine Schelbert. 2006. *Thinking Architecture*. 2nd, expanded ed ed. Basel ; Boston: Birkhäuser.

Chapter 8
Affective Opportunities and Recognition in Medical Education

Abstract In this final chapter, I will explore how students in medical schools can benefit from affective opportunities and recognition. To better understand the context of this debate, I will first of all take a socio-historical perspective primarily working from the post Second World War and exploring classic works in medical sociology to better understand how medical students were trained to separate affects from their future medical practices. I will then move to the present day to see how affects continue to be handled, dismissed, and unrecognized. Once I have shown the problem, as my ambition is to find promising methods to work toward affective opportunities and recognition in these cadres, the final part of the chapter will investigate some ideas to give a greater place for affects.

Keywords Medical students · Affects · Analysis of practice · Balint groups · Affective humility

8.1 Introduction

Thus far in my journey to find affective resources, opportunities, freedoms, and recognition for patients, healthcare providers, and families, I have looked at a variety of strategies that may be used in the hospital space and outside of it. This has included narrative methods, the elaboration of spiritual care, the integration of fellow patients into hospital policy and delivery, and the contribution of hospital design and architecture. However, I have thus far little discussed training and how it may instill a culture in which affects are disregarded. Without paying attention to the socialization processes that trains future healthcare professionals, it is unlikely we will be able to provide a safe affective space for them. Therefore, in this final chapter, I will take a step back to consider how we can better accompany medical students' affective needs as they learn to become doctors, in particular during their initial medical training and internships.

I have chosen to focus on medical training in particular. This is because, as we will see, those training medical students have been particularly ambivalent about the place

B. Bogaert, *Affective Injustice in Healthcare*, The International Library of Bioethics 112, https://doi.org/10.1007/978-3-031-94378-2_8

of affects and have sought to produce a separation between science-based practices and students' personal experiences and feelings. This has led affects to be disregarded or to be implicitly or explicitly discouraged in medical training. However, that is not to say that those in other types of professional training do not face these same kinds of challenges. For instance, while nursing-based pedagogies have long valued the relationship with the patient, including the emotions that may be lived by both actors in the relationship, the increasing turn to evidence-based practices is also having an impact on how affects are valued in training and practice. Furthermore, as we saw in Chapter 2, the hospital is one in which healthcare providers are encouraged to "put aside" their affects in order to work together and for the benefit of the patient. What we can see from these trends is that socialization processes both within training and outside it are having an impact on how affects are valued and recognized by healthcare professionals.

Following the analysis throughout the second part of the book, I will be inspired by Gallegos' (2021) conceptualization of the ways to work toward affective injustice, namely by providing affective resources, opportunities, and recognition. Medical students will both need the resources to be able to express their affects during the difficult medical socialization process, opportunities offered through classes and other peer groups, but also recognition by their peers and their professors that their affects are valid and say something important about care. As we will see through our socio-historical analysis in the first section, if these three conditions are not united, there is unlikely to be a change in how medical education is conceptualized or practiced. In a first section, I will therefore look at classic works of medical sociology to investigate how emotional socialization took place in medical training following the Second World War. I will then turn to the present day to determine whether or not affects have since then been given a greater place in medical training. Finally, I will introduce some methods that facilitate greater affective resources and opportunities for medical students.

8.2 Emotional Socialization in Medical Training and Practice

In this first section, I will look at the ideas that have led healthcare providers' affects to be silenced and devalued in both medical training and practice. As we will see, in the turn to science-based practices in the post-Second World War period, affects were largely put to the side and considered "inappropriate" in patient care. While these ideas are slowly changing, both in training (due to greater emphasis on empathy and medical humanities training in medical school) and without (notably due to the rise of patient movements we have explored in Chapter 6), a certain distrust toward affects has been hard to shake. I will therefore first take a socio-historical approach to better understand what happened in the post-war period, before returning to what is currently happening in medical education today.

To begin, in the twentieth century, clinicians started to view their profession as emerging from a long history of superstition and subjective judgment. In order to do so, they were guided by a new value: scientific objectivity (Halpern 2001, p. 17). This is the idea that scientific claims and methods should not be influenced by personal interest or value judgements, but rather by the scientific method based on facts. In the period following the Second World War, scientific objectivity was increasingly championed, leading to the promotion of science-based practice and a medicine increasingly reliant on evidence-based practices (Moses et al. 2023). By using the scientific method to apply and organize data to make healthcare decisions, clinicians began to radically change how they delivered healthcare, notably by developing and applying standardized tools and procedures for the diagnosis and treatment of disease. The rapid and significant advances made in fighting disease in this period were particularly seductive and lent support to these new science-based practices. In comparison to past medical practices, such as blood-letting, which were often ineffective in fighting diseases, new models showed quick and impressive results. In the post-war period, cures and treatments were found for polio, diphtheria, whooping cough, tuberculosis, schizophrenia, rheumatoid arthritis, and open-heart surgery and transplantation was introduced for the first time. Given these impressive achievements, science-based practices were seen as the solution to free people from fear of illness and improve the chronic disabilities of aging (Le Fanu 2002).

While these practices provided new frameworks to guide clinical care, they are only part of the decision-making process the clinician had to consider when in front of their individual patient, who has unique needs, requests, affects, and desires. Given their new scientific framework, clinicians hotly debated what kind of relationship healthcare providers should cultivate with their patients. Equipped with their new ways of thinking, they advocated that their emotions were detrimental to patient care, as it decreased or prevented them from maintaining the objectivity necessary to make sound medical decisions based upon evidence. In this environment, clinicians began to advocate that emotions were to avoided in their communication with patients. In these evolving discussions, the healthcare provider's feelings toward their patients began to be seen as suspect as it interfered with being a professional and providing quality care.

These ideas led to new ways of thinking about how to train future doctors. Medical sociologists (Parsons 1951; Becker et al. 2004; Fox 1998; Smith and Kleinman 1989) working in this period have helped to better understand how professionals became trained to regulate or suppress their affects in these new ways of practicing the art of medicine. Among the first, Talcott Parsons (1951, p. 212) notably suggested that "affective neutrality" become one of the five principles guiding the physician. By affective neutrality, he was referring to the expectation that the physician strives for and maintains *objectivity* in relation to patients. According to him, in order to maintain this objectivity, it was necessary for doctors not to become "emotionally involved." However, how did this work in practice? How could doctors be trained to practice affective neutrality?

Learning how to regulate and suppress one's emotions was a socialization process that began in medical school, a process that has been described in rich detail by Smith and Kleinman (1989) and Renée Fox's (1998) ethnographic studies of this period. These researchers showed that during medical training, students learned to develop what Renée Fox called "detached concern." This started in the first year of medical school, when medical students started working on human cadavers, which helped transform the human body into an object. Students learned to objectify the body by using clinical language and to see the body as an anatomical point of contact. By separating their interaction with a human body from the person itself, they began developing a distance (detached concern) that enabled them to interact with the human body in ways that would be taboo in their personal lives. As these first contacts with a human body were highly disturbing experiences, students coped by finding ways to control their emotions. This was both an *explicit* emotional regulation strategy engaged by them (as they believed that showing their emotions was unprofessional) but also *implicit*, as they learned by example from their fellow students and professors. In Fox's study, medical students they interviewed said they believed that cultivating detached concern was part of the experience and training needed to become a good professional. They clearly marked a distance between themselves at work (objective professionals) and their private lives (where their affects could be expressed and recognized).

However, that did not mean this emotional socialization process was an easy one for these students. Smith and Kleinman (1989) show how difficult it was for students to engage in these emotional regulation strategies. For instance, when feeling uncomfortable because they are making physical contact with persons in ways in which they would usually define as inappropriate in a personal context, the most common strategy engaged by them and their fellow students was to turn their attention to a specific body part in order to reduce the person to a mechanical or analytical problem. Another strategy used was to transform the problem into an intellectual puzzle (Smith and Kleinman 1989, p. 60). This helped them to distance themselves from the person they were working on and to focus on the body part, rather than on the disturbing emotions they were feeling based on what they were doing to the person. This strategy helped students to distance themselves from their affects, enabling them to work on the human body with which they were ill at ease. While this strategy was effective, it did little to deal with why they were feeling these things and what this may mean about their future interactions as doctors and the approach they would take with patients. At the same time, it was considered taboo to discuss such experiences in front of their peers. Therefore, these affects were put into the background and explicitly ignored.

This was the first step in the emotional socialization process. Fox (1979) shows how detached concern was next developed during the student's first experiences of an autopsy. Because students had already learned to emotionally detach themselves from the cadaver in the autonomy lab, it was believed they were now ready to approach those closer to the living. Fox shows that working on an autopsy was different than working on a cadaver, as the person was recently deceased and their medical chart was often in plain view. The color of the patient was even different: students said that while the cadaver was almost "not human" due to discoloration, the recently deceased person was closer to the real person in both color and touch, making it more difficult to dissociate it. Thanks to their experience with cadavers, they were able to use the same methods to emotionally distance themselves from this new situation. Fox, echoing Smith and Kleinman, showed in their study that students were able to emotionally detach themselves by focusing on the medical problem at hand (the pathology to be found, the techniques of the anatomist). The puzzle of the case in question also helped students to detach themselves from the situation and to "be objective" or "focus on the science" rather than the human body and the affects they were experiencing.

As already shown in the example on cadavers, while these individual strategies were largely effective, they were also helped by example. In these uncomfortable situations, they looked for guidance from the medical personnel in the autopsy room to show them how to act. By witnessing that professionals did not react emotionally during the autopsy—and similarly, the lack of emotional reaction in their fellow students—they saw several clear examples to emulate. Fox also shows that the clean, sterile, and impersonal room helped to distract students from their emotions. Just as in the cadaver room, the autopsy room became a means of practicing scientific objectivity, by "focusing on the science" rather than the person in front of them.

As seen from these examples, these medical students experienced intense emotions (disgust, fear) in these first training experiences, but also more vague feelings, such as discomfort and unease at the idea of touching dead bodies. In order to cope during these uncomfortable situations, they quickly developed emotional regulation strategies to get by, strategies that largely involved distracting themselves from what they were experiencing. These strategies included in particular distracting themselves from the situation (by focusing on the science), but also by emulating reactions from their fellow students and faculty (acting by example). While these strategies helped them to ease their comfort and focus on the training, it did little to acknowledge or recognize the emotions they were experiencing and the reasons for them. In turn, as Smith and Kleinman (1989) have shown, these feelings were little recognized by faculty, leading students to understand it was up to them to find ways to regulate their emotions. Students learned early on that they were not supposed to talk about their feelings with faculty members or other students, as it would be frowned upon or considered unprofessional. Students interviewed by Smith and Kleinman therefore used terms such as "screwing up your courage," "getting control of yourself" and "putting feelings aside" (1989, p. 59), from which we can see an emotional socialization process initiated by them but also encouraged by other students and professors. As discussed, these processes stemmed from these new scientific-based ideals of the doctor as an objective and rational scientist, free from emotions and other biases. Furthermore, they understood from observation that it was not until they learned to suppress their emotions that they could be taken seriously as medical professionals.

These first steps were part of a process to prepare them to see patients. First the autopsy, then the cadaver, and then finally they were ready to see live patients. Fox shows the detached concern they developed now could take on a new form, that of developing an equilibrium between showing attention and concern to the individual who was treated—but also being able to detach and objectify the sick body in order to treat the disease. This capability became the hallmark of the successful physician. According to Fox,

> The physician is required to strike a comparable balance with respect to the scope of his relationship and involvement with the patient. The physician is expected to confine his relations with the patient to matters which pertain to the handling of medical problems. However, he is also expected to establish a relationship with the patient which is comprehensive enough to enable him to know the 'patient as a person' and give substantial attention to (his) psychological and social circumstances (Fox 1979, p. 86).

What can be seen from this quotation is the expectation that the future professional will develop a "professional stance" in front of their patient, one that involves the distance necessary both for medical objectivity, but also one that centers the relationship on the medical problem at hand. From Fox's quotation, we can see that while the clinician is expected to get to know the patient, this is from a medical perspective: if it is to get to know the patient's biopsychosocial condition, it is for the purpose of finding the appropriate treatment and/or diagnosis of the disease. We see here again the cultivation of a professional distance in order to "objectively" treat a particular disease, one that takes into consideration the person and their specificities, but only in order to be good professionals and practice in line with evidence-based practices.

Today, the flaws in this way of conceiving of medicine—while largely continuing to be practiced—seem obvious. It ignores the singular colloquy between two human beings when confronted with each other in the consultation. It puts aside the affects that both may experience in this relationship and how they may both be affected by suffering. While they are emotional regulation strategies turned toward the patient, they do little to focus or pay attention on what the healthcare provider is experiencing or even what the patient is experiencing, beyond any factors outside of the immediate diagnosis and treatment.

Given their ways of seeing affects and the desire to practice and cultivate scientific objectivity, it is not surprising that medical schools in this period developed few resources to help students deal with their emotions. However, the problem is students were deeply emotionally affected by these early training experiences. Smith and Kleinman's study in particular make clear the anxiety medical students undergo in trying to adapt to their new roles, one in which they were largely left to deal with their affective experiences on their own. Smith and Kleinman also warn that these same strategies may also encourage them to dehumanize how they see the patient, notably by transforming the patient or procedure into an object, blaming or joking about patients, or avoiding contacts which troubled them, behaviors the researchers witnessed in their ethnographic studies (Smith and Kleinman 1989, p. 57).

The value of these early ethnographic studies is to show that emotional socialization processes in medical training are largely forms of emotional regulation, or a silencing of, hiding, or controlling affects based upon ingrained ideas that they are unprofessional or get in the way of patient care. As we saw from the examples given by the ethnographic studies, the strategies were cultivated by students themselves (based upon their own ideas), but also implicitly through example (by similar behaviors in professors and fellow students) from the very start of medical training. They were therefore influenced by their own ideas of what is a doctor, but they were also part of a socialization processes moved along by those both in hierarchical and similar positions. These studies also show the limits of these strategies by demonstrating how it may lead to a lack of concern for the patient or less quality care.

8.3 Is Detached Concern Still Relevant in Present day Medical Training?

Having elaborated these classical ethnographic studies in medical sociology, it will now be important to consider to what extent detached concern is still promoted in healthcare training and its institutions. A critique that may come from reading the section above could be simply that this was how medical training was in the past, but healthcare training takes better care of the emotional experiences of its students today. Let us therefore take a contemporary perspective to help nuance this debate and better understand if healthcare professionals' affects continue to be devalued in medical training today.

First of all, it must be recognized that around the same time as the idea of detached concern started to be promoted, other intellectual debates held competing views. Among these was the development of patient/person-centered care movements. Indeed, during the civil rights movements in the 1960s and 1970s in particular, ideas of greater patient participation came to the forefront and were advocated by citizens—but also by paramedical professionals, in particular nurses who advocated greater patient participation as part of their attitude and mission toward patients. This led to conceptualizations on shared decision-making (SDM). These new ideas came to a head in patient advocacy movements during the AIDS crisis in the 1990s, and since then the strong advocacy role of patient associations in chronic illness has helped move forward the idea that patients have a right to be actively engaged in care. These new models favored a greater "mutual investment" (Balint 1968) of healthcare providers in their relationships with patients. Under them, it became undesirable that a healthcare provider be a cold, detached professional. Having a relationship based on trust—a necessary condition for SDM—likely meant that each person had to give a little to the other, a part of themselves.

How did these care ideals affect medical students? Following up on her study, Fox showed that medical students in the 1960s and 1970s disapproved of their professors "all-knowing" and "omnipotent" attitudes and advocated that they should approach patients as "human beings." In practice, this meant showing respect for their feelings and opinions (Fox 1979, p. 100). It was in this way that Fox proposed they could now show "emotional and moral" integrity in their relationships with patients. This new disposition entailed being more honest and consistent in relationships with patients and being self-critical toward themselves and healthcare institutions. Fox says this new "medical student" disapproved of detached concern and placed value in feeling with the patient. When detachment was necessary, they did so with regret.

What these movements have accomplished is to call into question the ideal of the human body as an object. Patient/person-centered care movements in particular place emphasis on the individual person, on their affects, needs, and desires, on seeing the patient as a person in their singularity. They have also led to a greater place for empathy in medical curriculum. Jodi Halpern's book *From Detached Concern to Empathy* (2001) has notably advocated for this idea, arguing that cultivating empathy will help clinicians toward better communication with and understanding of their patients. Newer movements, such as the promotion of medical humanities training or Rita Charon's model of narrative medicine (Charon 2008)—have also sought to develop more empathetic capabilities in medical students.

In this environment, the patient's affects have been given greater recognition, but what about the healthcare provider's affects? There are two problems here. First of all, from Halpern's proposal to Charon's model and even in Fox's ideal of the new medical student—we witness a focus on the *patient* and their affective needs and concerns. Attention is turned toward the patient and their needs (a care that is patient-centered). Empathy toward the patient is cultivated as it will help improve clinical decision-making. Practically, this means taking seriously the patient's emotions, if not as form of knowledge, then at least as something legitimate in the patient's experience of illness. In other words, these models are turned toward the patient, rather than toward the healthcare professional. While doctors are expected to show concern toward their patients, this does not necessarily mean that their own affects are recognized or understood.

Therefore, we still have not resolved the problem of the affects that healthcare providers are experiencing in these encounters and what to do what them. We have not dealt with the continuing ambivalence of healthcare providers toward their affects and the implicit and explicit pressures they face in medical training and practice to learn to suppress them. This thinking is slowly changing, notably in evolving concepts of person-centered care (Ekman et al. 2021), where the emotional needs of the healthcare provider and the family are becoming more recognized alongside the patient. In these new models, greater attention to the mutual affects of each person is starting to emerge. However, at the current time, there is a still a long road ahead between theory and practice. The basic structure of medical training has not significantly changed, with early training based upon cadavers and training focused on science-based practices. While there is greater recognition of the affects students experience in these encounters, these remain exceptions rather than standard practice. Although there is increased awareness of the distress and burn-out faced by medical students, including those due to distressing experiences in medical training (Prendergast et al. 2024; Joshi et al. 2023; IsHak et al. 2013), there continues to be an insufficient space to discuss care experiences with faculty and peers.

However, one of the ways that has been increasingly promoted to improve medical curriculum to take into account the "human" behind the body to be treated and diagnosed is the integration of medical humanities training, notably courses taught by humanities scholars (literary, philosophy, history, sociology, and anthropology)—and increasingly by patient teachers. However, such training is still at the margin in most curriculums and many courses are adapted to fit with more concrete pedagogical objectives, such as improving communication strategies for future doctors. Furthermore, affects are not necessarily prioritized in these courses—and in some places actively discouraged by medical humanities scholars—to fit with other pedagogical goals and expectations. Therefore, medical humanities training in its current form has not (yet) changed the ambivalent place for affects.

To better understand how implicit and explicit emotional regulation strategies are still actively encouraged in medical training, let us take a common example that I have witnessed under the rubric of "medical humanities." A common methodology in early professional education, and in particular when students are starting their first internships, is to do role plays with fellow students. It is common to do role plays either as discussions with a patient or to practice interprofessional coordination meetings. While their explicit purpose is to help prepare students for future situations by giving practical tips of strategies, it also teaches them what teachers expect of future doctors. In other words, it tells them what is the right and wrong way to act. Today, these kinds of courses are both taught by doctors and by medical humanities scholars, sometimes separately and sometimes in duo. In these types of courses, it is still frequent for teachers (medical and humanities teachers) to stay to students in preamble, "here is not the time to express your emotions. Focus on the patient. Stay professional." Teachers that actively discourage affects in these situations do so with the intention of helping students take a professional posture toward other professionals and in their methodologies with the patient; however, they also (intentionally or unintentionally) are discouraging their affects and telling them they are unprofessional. This situation is therefore not substantially different from what the classic medical sociologists have shown about the emotional socialization processes they witnessed in the post-war period in medical training.

The difference in medical pedagogy today however is some teachers will choose to use to integrate a debriefing period following such role plays, which may enable these moments to become a safe space for affects. I will discuss in more detail this methodology in the next section. However, briefly, what we can see with this methodology is that expressing affects may still remain a difficult exercise for students. In the first place, they continue to be socialized to believe affects should be suppressed and they are unprofessional. In the second place, it is difficult to show and express these affects in front of students and professors for these reasons. Here the teacher can help overcome these barriers by clarifying that there will be time to discuss what they experienced during the debriefing period; however, in reality, these forums more often focus on the technical aspects of the case.

As seen from this discussion, the ambivalent place for affects in the medical curriculum has not substantially changed. In addition, while this period of medical training at the university is a pivotal one for students to understand ways to act, it is also reinforced by their early experiences in internships in hospitals. The lack of recognition and safe spaces for affects in hospitals tends to reinforce these ideas and puts them solidly into place for the young doctor. Clearly, we have a long road ahead if we are to take seriously healthcare providers' affects in training and practice.

8.4 Methodologies to Better Value and Recognize Affects in Medical Training

In this section, I will discuss several promising ways to give affects a greater place in medical training. I will focus in particular on three methods. The first will be the possibilities of different types of analysis of practice (Balint groups in particular) to serve as affective resources and opportunities; the second on how role play may be adapted to help provide a safer space for affects; finally, I will return to the discussion on affective humility and how to cultivate it in medical training. For this discussion, I have been particularly inspired by teachers and pedagogies that I have witnessed, studied, or experimented with in humanities teaching in medical schools in France and Switzerland. I have chosen these methodologies to discuss because, while they are not necessarily "new," they already exist and work. This means they can easily be adapted and integrated to pave the way for greater affective justice in these cadres. However, it must be said that while some of the pedagogies are already in place in some contexts, in others they remain rare and underdeveloped. For instance, the first is the so-called Balint groups, which give a greater place for recognition of affects among peers. While they are increasingly being promoted in English-speaking contexts in medical schools (often at the initiative of students themselves), they are rare/non-existent in the French-speaking context where I work (France, Switzerland), although as I will show, I was able to propose and integrate this pedagogy into some aspects of the curriculum. Therefore, it is not be assumed that these types of classes are solidly in place in medical training, but rather that they represent a few promising initiatives that merit wider implementation to work toward affective justice in these cadres. In this section, I will discuss the possibilities, limits, and adaptations that may be needed for them to serve fully as affective resources and opportunities. Table 8.1 summarizes the main themes to be discussed.

As seen from the above table, I will come back to a number of discussions that have reoccurred in this book, in particular the integration of fellow patients and alternative pedagogies such as narrative medicine. However, I will first focus on methodologies that already widely exist in medical curriculums to better understand how they may be adapted to work toward affective justice.

8.4.1 Balint Groups and Other Forms of Analysis of Practice

In this first section, I will discuss the methodology of analysis of practice and how it may become an affective resource for future doctors. By analysis of practice, I am referring to a global term that incorporates various ways in which students or young doctors can discuss conflictual or difficult situations in a safe space among peers. In these discussions, as we will see, affects have a strong place and can be recognized and taken seriously as a form of knowledge. While many forms of analysis of practice exist, I will discuss in particular Balint groups; however, it is be acknowledged that

Table 8.1 Pedagogies to work for affective justice in medical training

Type of pedagogy	Relation to affective justice	Limits
Analysis of practice	Provides affective resources and opportunities for students to discuss difficult emotions	Their ability to provide safe spaces depends on the teacher and environment created by their peers
Role plays	Enables affective opportunities for students to openly voice affects and their relationship with care values	
Patient teachers	Allows patient affects to be considered forms of knowledge (patients become epistemic authorities), encouraging epistemic and affective humility	Choice in teachers influences to what extent humility may be cultivated (need for a wide diversity of teachers to cultivate multiple perspectives)
Narrative medicine	Narrative humility can be cultivated (a recognition of the other as not fully understandable)	Has a marginal place in the curriculum at the current time; will depend on the teacher/facilitator and their abilities to cultivate humility in students

many other informal and formal types of analysis of practice exist in both medical training (Gishen et al. 2016) and in hospitals to promote a safe space for affects (Schwartz et al. 2020; Taylor et al. 2018; Farr and Barker 2017; Gardner and Bray 2014), as already discussed in this book. The most common of these are the so-called Schwartz rounds, which are generally much bigger and rely less on the contribution of each members, but may provide the kinds of affective resources and opportunities we are looking for, in particular by showing students it is normal and acceptable to experience affects toward patients and in difficult care situations (Gishen et al. 2016). However, I have chosen to elaborate the methodology of Balint groups, as they rely on a small group of students and involve more active discussion, which is necessary for those students who are early on in their medical training and need to discuss openly their internship experiences with their peers.

To start the discussion, Balint groups were first developed by George Balint and his wife Enid in the afterwar period for groups of general practioners to discuss difficult patients and cases. In their original form, they were facilitated by an expert in psychoanalysis, who provided an outsider's view to guide discussions (Reznik 2009). The aim was to discuss relationships between the doctor and the patient, and the facilitators encouraged doctors to retell cases in free form, including in contradiction to their case notes. This was done to allow a complete picture to emerge of the situation, beyond the "cold case file" and all that was absent in this form of reporting. According to Balint (1985), their work was based upon the idea that doctors unconsciously defend themselves against certain thoughts and ideas. This was because they attempted to structure these experiences, but it also led them to leave out their facts and feelings about them. The idea of free form telling was to allow "other information" to come forth without embarrassment. According to them,

> We run our groups the way we do so doctors can be active, not passively receptive, either of their own feelings or what the leader says; so they can talk freely about their patients

> and their feelings about them…in so doing they can get in touch with feelings in themselves about which they have been unaware and which may enable them in due course to understand something about their patients that they would not have been able to do had they been out of touch with their own feelings and the seriousness of them (Balint 1985, p. 8-9).

As seen from these descriptions by Enid Balint, even in their original form these groups were designed to bring affects to the forefront, to take them seriously as forms of knowledge, and to investigate what they had to say about care and relationships with patients. According to them, by embracing these other facts, ideas, and experiences, the methodology provided a gateway to new facets of patient experience and of the doctor-patient relationship. It gave them consciousness of these other facts/experiences—including how they felt about the situation and the patient—and in this way, helped them to better understand the patient's perspective. It led to creation of a safe space enabling them to share unique facts of the cases and to normalize feelings and emotions experienced, including those particularly intense or negative (Farkas et al. 2024). Balint groups may therefore provide a number of affective resources and opportunities: (1) provision of a safe space for discussion of the doctor's feelings (affective resources); (2) affects are actively searched for and given meaning by the facilitators (affective opportunities); (3) doctors are able to have these affects valued and recognized by their peers (affective recognition). This method permits affects both to be seen as forms of knowledge and a way to improve care and the doctor-patient relationship, significantly changing how they are viewed and treated in medical practice.

Given these possibilities, let us now explore how such groups can be integrated into medical training, in particular when students have limited experience as healthcare providers and often take an observer role in their early internships. In medical schools, it is not always feasible or practicable for these groups to be led by an expert in psychoanalysis. Balint groups have also been questioned for their psychoanalytic emphasis and its limitations, which suggests that the integration of this type of expertise is not necessarily desirable in all circumstances, and in particular for medical students who may not adhere or fully capture the implications of this type of facilitator. I will therefore discuss how to adapt these kinds of groups without psychoanalysis, but in keeping with the ambition of providing affective resources, opportunities, and recognition in the same manner.

These groups have been reimagined differently for medical students, in particular because students focus on situations they have lived through in their internships. In these early experiences, medical students are observers, and they often experience care in a passive way (Torppa et al. 2008). This does not mean however that they do have opinions, experiences, and strong impressions of these experiences: in their position as an observer, and in their early medical training, they are often surprised, shocked, and confused about what they have observed, and in particular the disconnect between their ideals of what a doctor should be, how they should act, and what they have observed in their interactions with patients. This temporality of medical training therefore seems ideal to create such types of affective opportunities to explore students' experiences and how it may connect to their care ideals. By providing affective safe spaces, by focusing on the whole internship experience (and not just the technical aspects of the case), they encourage critical reflection and questioning and mutual learning among students. When they organized by students among themselves, they facilitate affective expression and recognition among peers; when they involve a teacher, this may also change how the student views affects and their place in the medical curriculum.

However, the problem with this methodology as it is currently practiced is that it is largely done on a voluntary basis outside of more formal classes. This means those who are most reflexive, who have the most questions, or who are more willing to discuss their feelings may join these groups. It is therefore worth seeing to what extent they may be integrated into more formal classes so that all students may benefit. I will here discuss a practical example of how I have integrated the Balint curriculum (with adaptation) into a more formal class structure. It is to be noted that my example was in class that was part of a medical humanities curriculum in France for those medical students interested in ethics and social and human sciences; thus it was already a group open to such alternatives ways of caring. As I have been advocating, it would be more fruitful to integrate such methodologies with a wider group of students, rather than having them be merely "optional" in the medical curriculum to enable a wider student population to benefit. However, I will present my method here and its possibilities as an affective resource in a formal class setting. I will also present how the students evaluated the class.[1]

In inspiration of the Balint group methodology, the objective of the class was to change how students viewed affects as a whole, and in particular in relation to their practices. My goal was for them to understand them as a form of knowledge that may help them to improve care for their future patients. Therefore, as a first step, it was crucial to inform students about changing ideas on emotions. I therefore started the class by discussing recent studies in neuroscience, social sciences and philosophy on emotions. I discussed in particular Antonio Damasio's early work, which showed the link between emotions and rationality (1994). As he showed, without emotions, we would not be able to make sound decisions, adapted to the environment. I then brought forth philosophical ideas on emotions, and in particular Martha Nussbaum's ideas on emotions as intelligent and related to our values (2001). This helped in particular to show how so-called negative emotions (such as fear and disgust experienced in these first internship experiences), might show something valuable about their care experiences. This first step permitted students to understand that emotions were not "irrational" and could be linked to good decision-making in their future practice. It also showed that affects are connected to our values and what is particularly important to us in our lives.

Having paved the way to understand the value of affects, the next step involved a practical example of research on emotions in the doctor-patient relationship. In this step, I talked about one of the controversial results of my own research, in particular

[1] This case study was originally published in French in a collective work on field philosophy. The article discussed the restitution of research results to other types of stakeholders; therefore the analysis in terms of providing affective resources and opportunities for medical students is quite different. For the original article, see: Bogaert (2023).

an empirical research project which showed the fear experienced by patients and doctors in the face of medical uncertainty (Bogaert 2020). A point of discussion was the analysis by patients that their doctors were "scared" of them as they didn't know how they could help them due to the limits of medical knowledge. According to them, this fear of the patient led to hiding behaviors (looking at the screen instead of them, avoiding uncomfortable situations), which led to ineffective communication and shared decision-making. On the other hand, doctors interviewed talked to the researcher about apprehension or anxiety in the face of announcing bad news and stated that "fear" was too strong an emotion to express what they were feeling. Therefore, while both actors admitted that affects were an important part of the situation, they disagreed that "fear" was an accurate evaluation of the situation. This controversial result was therefore proposed as an open question to the class, notably to debate if this evaluation by patients was or could be accurate, but also what it might say about the doctor-patient relationship. In this debate, students were interested in the reasons why doctors could be reluctant to speak of their emotions or even to accept them as valid. As we can see, in this step students have already accepted affects as important and start to question why they are taboo in medical practice.

Having prepared students to accept their affects as forms of knowledge, I could now move on to the reformulated Balint methodology. I invited students to discuss in small groups cases that made an impact on them during their internships, including what troubled or interested them about these cases. Students were invited (without hierarchy or intervention by the teacher) to exchange and debate these case studies and how it impacted care. While the workshop was designed as a safe space through discussions in small groups, they were equally invited to discuss them in front of the class to allow others to benefit from their reflections. What I observed from these interactions is that students had a lot of material to discuss, as their experiences were rich in affects, both from the healthcare perspective as well as the patient one. Some discussed difficult emotions and situations, such as the fear they felt in announcing to a patient bad news. Others discussed how emotions motivated them in their future professionals, such as the joy experienced by the healthcare team during childbirth. They also discussed care strategies and patient-doctor relationships they wished to emulate or avoid based upon these observations.

The reformulated Balint methodology allowed a safe space for discussion of affects among peers. It showed them their importance in their future practice, but also the value of discussing them and helping each other understand and appropriate these experiences. Several students talked to me after the class to ask how they could organize this type of discussion and peer support at the faculty, either as a formal part of their class or in their free time. While they were free to organize such groups amongst themselves, they actively sought faculty support and encouragement in the creation and facilitation of such groups, suggesting that they wished it to become a formal part of the curriculum.

This was an example of a constructivist methodology that worked well to help students see and understand affects in a different light. Such pedagogies are more and more used in medical schools to change the way students interact with knowledge, their teachers and fellow students (Hennrikus et al. 2020). Constructivist pedagogies, centered on the student, are effective to help students develop adaptable beliefs and dispositions and critical thinking skills, in particular compared with lectures that engage patients in a passive way (Greene and Yu 2016). In this type of pedagogy, the teacher guides the students in the construction and evaluation of their own knowledge and that of others, and then helps students to take the initiative and explore and resolve problems with the help of their peers. The Balint methodology provided this kind of opportunity through peer support and an active discussion and questioning of both research results (the evaluation of fear by patients) and their own critical understanding and appropriation of their internship experiences. By helping students develop adaptable beliefs and dispositions, in particular through learning with their peers, these students started to see the value of their affects and their importance to their future practices.

Let us now discuss another promising methodology that may be adapted to provide greater affective opportunities and resources, namely the role play methodology, which already forms an important component of early medical curriculum.

8.4.2 Role Plays: Announcing a Diagnosis

Another methodology that can serve as an affective resource for patients is role plays. These forms of pedagogy are frequently used in medical schools to enable students to learn how to communicate with future patients, their families, and their colleagues. The idea is to train them so that they will be more at ease and prepared when faced with the real patient or colleague. They can be fully scripted (students act according to verbatim scripts) or partially scripted (in which certain prompts are given or factual information provided concerning a case). They may be designed to practice communication with a patient and their family, but they may also be designed to benefit from and understand the importance of interdisciplinary communication. Actors often rotate through roles, with the intention of gaining insights into other perspectives, from that of the doctor, patient, family member, or observer (Nestel and Tierney 2007). While there is need for more studies to determine the impact of role plays in medical education, preliminary studies show they may help medical students to develop greater empathy (Singh et al. 2017), in particular through experiential knowledge of playing a patient or family member. They have also shown it may help them to work through their anxieties, fears, and doubts (Ong et al. 2022).

There are several ways these pedagogies may serve as affective resources, opportunities, and recognition for medical students. This may happen in several temporalities, both before or after the role play itself. The teacher may for example commence the class by discussing difficult situations they have experienced in order to prepare for the role play. This may include a free discussion on how they felt about their first experiences, including their frustrations, problems, and fears. In this way, they will provide an affective safe space for students to discuss their real experiences and observations, similar to the Balint group. More commonly, they involve a debriefing period after the role play, in which affects and other information may be openly discussed to understand what went right or wrong during the situation (and potentially how it may have affected them).

The danger for our subject is that, as I have already discussed, this pedagogy may serve as an unsafe affective space depending on how affects are understood and treated by their peers and teachers. For instance, if they are told by the teacher "not to act with their emotions" at the start of the role play, this may encourage affective silencing. The other way it may serve as an unsafe space is if their peers do not recognize, take seriously, or allow affects to enter into the discussion, which will also encourage affective silencing of the student who may need this forum to work through and talk about their affects. As role playing in medical students also serves a socialization function, they are also the means in which future doctors learn what is the "right" and the "wrong" way to communicate with the patient. However, this is not inevitable, particularly when the pedagogy and the teacher explicitly encourage and give a space for affects. To better understand how such pedagogies may serve as an affective resource, I will rely on my observation and participation in role plays for third year medical students in France. The purpose of these role plays was to build their skills in announcing a diagnosis. These classes were co-taught by a psychologist and an ethics/social sciences teacher, although conceivably they could be taught by other types of faculty. However, the benefit, as we will see, of having a psychologist as the teacher was an explicit attention to and communication about affects.

In the situation I will describe, students were at a stage of their medical education in which they are starting to change from observers to practioners. When in internships, they are increasingly being given healthcare tasks, including responsibilities to discuss difficult situations—such as a diagnosis—with patients and their families. While technically the more senior staff member would be present and lead such communication, this was sometimes tasked to students, who often have to practice communicating a diagnosis for the first time to a worried patient or family member, with or without the help of more senior staff. In other situations, the senior staff was present and took the lead, and the student fell into the observer role. This gave them the opportunity to pay greater attention to the methods of communication with the patient and the family and to understand how the doctor's stance affected them. Students in this phase of sometimes observer, sometimes practioner, were in a steep learning curve. They needed to be prepared to communicate with the patient and their family about a diagnosis, and they were actively looking for clues of how to do so in ways that were both professional and respectful of the patient and their family.

It was considered important at this stage of their education be to able to practice telling a diagnosis, but also being able to discuss these experiences with their peers and with a psychologist, who could help them to work through difficult emotional experiences. In the class that I participated in as a co-teacher, the methodology used was to: (1) organize a roundtable with students who wished to talk about their experiences of announcing a diagnosis; (2) practice role-playing; (3) introduce a debriefing period to discuss what happened during the role plays. While the explicit purpose of such a pedagogy was not necessarily to give a voice and recognition to affects, they quickly became part of the discussion and even the main focus of the discussion given the uncomfortable situations students experienced.

In the first part of the class, the psychologist encouraged each person around the room to discuss their diagnosis experiences. Students during this period of their medical education were struggling with how they impacted patients through their role as a doctor and how patients reacted to bad news. Both from the standpoint of the future doctor and from the perspective of the patient and the family, there was considerable fear, anxiety, and stress surrounding these encounters in which the news they had to communicate to patients and their families was often devasting. Thus, in the first part of the class, it was necessary to give a safe space for students to openly discuss and express what they had lived during these early experiences, which was often very difficult and emotionally charged. The psychologist gave voice to, encouraged, and recognized their affects as important and helped them discuss through these first experiences and what it may mean about their future roles as doctors. Students felt a lot during these early experiences; they often cried in the retelling of them. Because the psychologist provided a safe space, some students were able to overtly tell their fears and anxieties in front of the faculty and fellow students.

Having gone through this process, the next step of the pedagogy was to practice role plays in announcing a diagnosis with fellow students. The purpose of this was to provide a safe space to practice and make mistakes, as well as to learn good practices and techniques from their peers. Here, the psychologist provided an additional affective opportunity by organizing a debriefing period at the end of the role play. While students could announce more technical aspects of the case, given the safe space promoted in the first step, it also became a forum in which they could return to their uncertainties, their feelings, their worries about these aspects of patient care.

This kind of pedagogy is both a simple and effective means of giving value to affects in difficult care situations. By discussing with peers, the students also understood that they were not alone in these difficult situations and the affects they experienced were common and could say something important about what kind of doctor they wished to be. This methodology functions in a similar way to Balint groups, in that affects are recognized and understood both by the teacher and their peers, and their affective expression encouraged through alternative discussions of real-life and fictional cases. In other words, it is not only the case files (with the medical problem) that is discussed, but rather the person-to-person relationship and how they are affected by these experiences.

The limitation of this methodology is not all students will want or wish to openly discuss their affects in such an environment. It will also involve an active commitment by the teacher—and fellow students—to provide that safe space. However, as we can see, by introducing the first phase, the teacher both actively promoted and encouraged their expression. For those who did not feel comfortable talking, it at least allowed them to understand these affects were an inevitable part of the care experience and they should not be ashamed to feel them (even if they are unable to explicitly discuss them in front of the teacher or fellow students).

To be clear, while I see a potential for role playing to serve as affective resources and opportunities for medical students, particularly when led by teachers trained to recognize and encourage their expression, that is not to say all role plays can or should serve this purpose. They are often used as pedagogical tools for early students to learn technical and communication skills before internships; thus, they need not always focus on such aspects. However, it will be vital to include a safe space for affects in at least some of them, in particular for difficult care situations such as announcing a diagnosis, which will inevitably make future doctors feel and question their roles and how to do well for their patients.

8.4.3 Developing Epistemic and Affective Humility in Medical Students

Having explored two promising ways that may be adapted to provide a greater space for healthcare providers' affects, in this final section, it will be important to return to patients and their affective needs. In this discussion, I will discuss emerging pedagogies that may help develop greater affective humility in medical students. By affective humility, I refer to an awareness that we do not necessarily experience and express the world in the same ways and an active disposition to look for and understand the other's affective experiences. In this endeavor, there is a recognition that it is not fully possible to have access to another's affects, as each person is different from the other given their individual histories and particularities. However, as advocated in this book, an attention to and development of affective (and other forms of epistemic) humility will encourage the person to be modest in their judgements and opinions, to actively seek to better understand the other, all the while fully recognizing they cannot fully have access to the other's experiences. In this way, developing humility may serve future doctors both in their relationships with their patients and families, but also to be more attentive to and understanding of their colleagues.

At the current time, humility is a subject that merits more research and pedagogical attention to improve care and care relationships. Research thus far has shown how humility can develop the doctor's *inward perspective*, including their views of their abilities and limits. It also encourages an intellectual openness to new knowledge, as well as widens their *outward perspective*, including an understanding and appreciation for the wider context in which they work and a greater value for the individual patient's experience. In this way, humility can positively influence care, encourage healthcare providers' motivation to excel in their professions by seeking out new knowledge, and improve relationships with colleagues and patients by being more attentive to the person's individual needs and experiences (Wadhwa and Mahant 2022; Huynh and Dicke-Bohmann 2020).

It is vital to start educating future doctors to develop both epistemic and affective humility at all stages of medical education to prepare them to become compassionate caregivers. Despite the advocacy and increasing research attention that has shown the importance of developing humility in medicine, at the current time it is not promoted in medical training, in particular in competency frameworks (Wadhwa and Mahant 2022). While many methodologies may be explored of how to develop epistemic and affective humility in medical students, in order to make the link with our previous discussions in Part II of this book, I will look at several methodologies already discussed—in particular patient participation in medical school education and the use of narratives—to explore their potentials and limits for this subject. Following Wadhwa and Mahant, I will advocate that integrating humility into core competency frameworks and the overall medical curriculum is the key to making this ambition a reality in future medical training and practice. However, in the meantime, it could already be happening in practice (at least partially and indirectly) through patient-teacher integration and the use of narrative medicine pedagogies, as I will now discuss.

8.4.3.1 Patient-Teacher Integration in Medical Schools

In Chapter 6, I looked the various ways in which the active participation of patient partners/advocates works toward greater affective justice for patients. In this section, let us explore the increasing integration of patient teachers in medical schools and the ways in which they may help future doctors develop affective humility. Patient involvement in medical training dates back to the 1960s, when they were present during mock physical examinations (Towle et al. 2010). However, their engagement has increased exponentially in recent years, from being involved in design of the curriculum to serving as patient teachers in the classroom and mentoring to individual students. While methodologies vary across contexts, patient teachers frequently intervene to give students feedback on history taking, role plays, and doctor-patient communication. They also regularly share their experiences in healthcare to enable students to better understand the patient perspective and wider experiences of living with an illness beyond medical aspects. Medical faculty are increasingly seeing the value of linking students with patients to improve student communication skills and gain insights into life with an illness outside of more formal, biomedical based classes (Stijntje et al. 2020).

There a number of ways in which patient teachers may facilitate epistemic and affective humility in future doctors. In the first place, by taking on the role of teachers and evaluators, they are showing that patients (and not only faculty) can be considered epistemic authorities about an illness, meaning their testimony will be important to take into consideration (and not just medical facts about a case). Through patient teachers, students can start to develop an outward perspective, one turned to valuing another's experiences, which is considered a key component of humility (Wadhwa and Mahant 2022). They can do so by listening, paying attention to, and valuing the telling of an illness, helping them to see the wider context and its effects on the individual person. As I have been advocating all along in this book, affects are a form of knowledge. By explicitly integrating them into the curriculum, the patient's affects—be it their fears, worries, anxieties, or joys—are given a legitimate voice and space and are considered a source of knowledge important for patient care. By hearing the patient's story, by understanding they may in the future have medical knowledge to give, we will be encouraging them toward such practices in the future. However, the value of this telling is also to help students recognize they will never fully be able to understand all of the affects the patient may be experiencing. In a patient's storytelling, affects hold a special place. They are singular; they are remarkable. They help show students the individual nature of the illness experience—for that particular person. The retelling of such illness experiences through affects are also those most likely to be remembered by the student when thinking about the class afterward and when they think about that particular patient. In other words, affects give that particular experience a force and a singularity that a more neutral, cold telling (such as through a case file) would be unable to achieve.

In this way, both by becoming epistemic authorities (teachers in front of students) and by giving access to singular accounts of experience, the integration of patient teachers in medical schools may help students to develop epistemic and affective humility. By seeing another perspective, by understanding the singular nature of the illness experience and the individual path each patient may take, the integration of patient teachers can help work toward this goal. However, this kind of pedagogy will only be successful if combined with other methodologies—including a demonstration of humility by their faculty and peers—to give an example for future doctors to emulate and for it to become a virtue in medical training. There will also be a need to integrate the experiences of other healthcare professionals—nurses, care assistants, physical therapists—to enable them to become the epistemic authorities in telling about their own professions and encourage this kind of humility toward other care professionals.

However, before advocating the integration of patient teachers for these reasons, I need to point out a risk. As seen in Chapter 6, while ideally patient teachers should come from a variety of backgrounds, there is currently an overrepresentation of highly educated groups. Another problem is the integration of those patients who are particularly knowledgeable of their illness, who have found a successful treatment or cure, and in particular those with agree with their medical treatment plan and have collaborated successfully with doctors. Some may also have professional patient partner training, which may encourage them to not express their affects, including in their retelling of their care experiences, in order to stay more objective/professional. While these persons represent a tiny minority of the patient population, they are the majority of those who teach in medical schools. The problem with this lack of diversity is that it will promote a certain vision of the patient and to what standards, ideals, and means of interacting are considered ideal. In this way, it harmonizes with—rather than promotes healthy argument—about patient needs and priorities and often aligns with biomedical understandings and the information that other faculty members wish to put forward. Given the whole purpose of integrating these persons is to bring forth a new perspective, this lack of diversity may go against the goal of promoting epistemic and affective humility. This may happen in particular when these persons decide to stifle or suppress certain negative emotions they have felt in care, or because they themselves have had good experiences and relationships with healthcare providers (particularly when they come from higher socio-economic groups). In order for students to be able to see and understand the limits of their knowledge, to encourage them to actively seek out alternative views and cultivate this openness, they will need alternative, conflictual, and in particular a diversity of perspectives from patient teachers. This means finding ways to include vulnerable groups—in particular those that have been marginalized or had conflictual relationships with healthcare providers, in order to provide a balanced view for students. Now, having explored the potential of patient teachers to develop humility, let us now turn to a second methodology, namely the use of narrative medicine in training.

8.4.3.2 Narrative Humility

In this final section, let us return to the question of narrative medicine pedagogies (Chapter 4) and how it may help students cultivate greater epistemic and affective humility. As discussed, while such pedagogies continue to have a marginal place in medical school today, with many courses falling into the "optional" category, they provide particularly interesting and innovative ways of working toward epistemic and affective humility of medical students. Indeed, narrative medicine researchers have in particular been interested in how such pedagogies may work toward greater humility. While Rita Charon has advocated that clinicians develop narrative competence, DasGupta (2008) suggests that witnessing, reading, and discussing stories of suffering will enable the development of *narrative humility*.

In the first place, reading (or hearing) a patient's story, as we have seen in the proceeding section on patient teachers, means that for the listener or the reader, their story belongs to them. Citing Irvine (2005), DasGupta (2008) says narrative humility is a recognition "toward that which we do not know—the face of the Other, the face we cannot know but to which we are responsible." Irving shows literature points medicine beyond its own reflection, in particular by giving the reader another form of representation, another way of seeing (Irvine 2005). This leads to an acknowledgement that patient stories are not objects to be mastered, but rather ones that can be engaged with, ones in which we can question our expectations, reactions, emotions felt toward the story, including whether it attracts or repels us because of our personal experiences and beliefs. As DasGupta says,

> Narrative humility allows clinicians to recognize that each story we hear holds elements that are unfamiliar—be they cultural, socioeconomic, sexual, religious, or idiosyncratically personal. Assuming that our reading of any patient's story is the definitive interpretation of that story is to risk closing ourselves off to its most valuable nuances and particularities.

As can be seen from DasGupta's conceptualization, integrating narrative medicine pedagogies in medical education can therefore cultivate the kind of epistemic and affective humility that we are seeking. As discussed in Chapter 4, the patient's and doctor's stories are full of affects. They make the story vivid, remarkable, even unforgettable—all the while pointing to what is important to the writer or character. Narrative medicine workshops, by their pedagogy of close reading and discussion, cultivate critical thinking about how the story is told and the reader's reaction and engagement with the story. They provide an insider view in what it means to live that experience—but also a recognition that the reader will never fully understand it or become the other. In this way, the telling of illness experiences by patient teachers also serves this purpose, particularly when students can directly ask questions of, challenge, and try to better understand the patient's story. The time taken to see the person in their globality, the opportunity to hear and understand their story, is a key facilitator in cultivating humility—and therefore in working toward affective and epistemic justice toward future patients. The capability to promote such narrative humility will however need to be made explicit by the teacher. If they only involve a discussion of the character or the plot, rather than an active questioning of the person's reactions to and engagement with the text and its characters, or engaging an active debate about their capacities to fully understand and emphasize with them, then a recognition of affects and the development of greater humility will likely remain in the background.

These are just a few ways in which epistemic and affective humility may be cultivated in medical students. Should it be given an explicit place, in particular as a core competency to be developed for medical students, it is likely to developed further, in particular through integration into other medical pedagogies, but also to benefit from increased recognition and commitment by other faculty members. Teaching students to be humble in their approach will help cultivate both inward and outward perspectives, to be better professionals and build better relationships with patients and colleagues. It therefore needs to be understood as a core competency in medical training.

8.5 Conclusion

Given that medical school is where future doctors are socialized and trained, they are both the ideal starting and ending place for our discussion on affective justice in healthcare. This chapter helped to better understand how medical schools may overcome the emotional socialization processes that leads to affective injustice for patients, families, and healthcare professionals. As seen throughout our historical and current analysis, this is no easy feat, but is slowly starting to change with a more active recognition of the importance of affects in medical curriculum and methodologies that give greater space for their expression. As this discussion has shown, an open space for affects can be easily integrated into existing courses and priorities, in particular if teachers are committed to providing affective opportunities and recognition for their students. The integration of narrative methodologies and of a diverse profile of patient teachers is also an important means of cultivating epistemic and affective humility, which will enable future medical professionals to become compassionate carers.

References

Balint, Michael. 1968. *The doctor, his patient and the illness.* 2nd ed., revised&enlarged, reprinted. London: Pitman Paperbacks.

Balint, Enid. 1985. "The History of Training and Research in Balint Groups." *Psychoanalytic Psychotherapy* 1 (2): 1–9. https://doi.org/10.1080/02668738500700111

Becker, Howard S., Blanche Geer, Everett C. Hughes, and Anselm L. Strauss. 2004. *Boys in White: Student Culture in Medical School.* Fac-sim Ed. New Brunswick London: Transaction.

Bogaert, Brenda. 2023. "La restitution d'un travail en philosophie de terrain : pour qui, pourquoi, et comment ?" In *Manifeste pour une philosophie de terrain.* Philosophie de terrain. Dijon: Philosophie de terrain.

Bogaert, Brenda. 2020. "Untangling Fear and Eudaimonia in the Healthcare Provider-Patient Relationship." *Medicine, Health Care and Philosophy* 23 (3): 457–469. https://doi.org/10.1007/s11019-020-09956-1

Charon, Rita. 2008. *Narrative Medicine: Honoring the Stories of Illness.* 1. paperback ed. Oxford: Oxford Univ. Press.

Damasio, Antonio R. 1994. *Descartes' Error: Emotion, Reason, and the Human Brain.* New York: Putnam.

DasGupta, Sayantani. 2008. "Narrative humility". *The Lancet* 371: 980–981. https://doi.org/10.1016/S0140-6736(08)60440-7.

Ekman, Inger, Zahra Ebrahimi, and Patricia Olaya Contreras. 2021. "Person-Centred Care: Looking Back, Looking Forward." *European Journal of Cardiovascular Nursing* 20 (2): 93–95. https://doi.org/10.1093/eurjcn/zvaa025

Farkas, Kinga, Gábor Csukly, and Peter Fonagy. 2024. "Is the Balint Group an Opportunity to Mentalize?" *British Journal of Psychotherapy* 40 (1): 55–75. https://doi.org/10.1111/bjp.12880

Farr, Michelle, and Rhiannon Barker. 2017. "Can Staff Be Supported to Deliver Compassionate Care Through Implementing Schwartz Rounds in Community and Mental Health Services?" *Qualitative Health Research* 27 (11): 1652–1663. https://doi.org/10.1177/1049732317702101

Fox, Renée C. 1998. *Experiment perilous: physicians and patients facing the unknown.* New Brunswick, N.J., U.S.A: Transaction Publishers.

Fox, Renée C. 1979. *Essays in Medical Sociology: Journeys into the Field.* Health, Medicine, and Society. New York: Wiley & Sons.

Gallegos, Francisco. 2021. "Affective Injustice and Fundamental Affective Goods." *Journal of Social Philosophy* n/a (n/a). https://doi.org/10.1111/josp.12428

Gardner, Sharryn, and Dominic Bray. 2014. "Compassionate Conversations." *BMJ Quality Improvement Reports* 3 (1): u204059.w2102. https://doi.org/10.1136/bmjquality.u204059.w2102

Gishen, Faye, Sophia Whitman, Deborah Gill, Rhiannon Barker, and Steven Walker. 2016. "Schwartz Centre Rounds: A New Initiative in the Undergraduate Curriculum—What Do Medical Students Think?" *BMC Medical Education* 16 (1): 246. https://doi.org/10.1186/s12909-016-0762-6

Greene, Jeffrey A., and Seung B. Yu. 2016. "Educating Critical Thinkers: The Role of Epistemic Cognition." *Policy Insights from the Behavioral and Brain Sciences* 3 (1): 45–53. https://doi.org/10.1177/2372732215622223

Halpern, Jodi. 2001. *From Detached Concern to Empathy: Humanizing Medical Practice.* Oxford: Oxford Univ. Press.

Hennrikus, Eileen F., Michael P. Skolka, and Nicholas Hennrikus. 2020. "Social Constructivism in Medical School Where Students Become Patients with Dietary Restrictions." *Advances in Medical Education and Practice* Volume 11 (July): 505–511. https://doi.org/10.2147/AMEP.S259727

Huynh, Ho Phi, and Amy Dicke-Bohmann. 2020. "Humble Doctors, Healthy Patients? Exploring the Relationships between Clinician Humility and Patient Satisfaction, Trust, and Health Status." *Patient Education and Counseling* 103 (1): 173–179. https://doi.org/10.1016/j.pec.2019.07.022

Irvine, Craig. 2005. "The Other Side of Silence: Levinas, Medicine, and Literature." *Literature and Medicine* 24 (1): 8–18. https://doi.org/10.1353/lm.2005.0027

IsHak, Waguih, Rose Nikravesh, Sara Lederer, Robert Perry, Dotun Ogunyemi, and Carol Bernstein. 2013. "Burnout in Medical Students: A Systematic Review." *The Clinical Teacher* 10 (4): 242–245. https://doi.org/10.1111/tct.12014

Joshi, Vivek R, Jimmy Michael Younger, Sanjib Das, B.K. Manjunatha Goud, and Kartick Pramanik. 2023. "Factors Influencing Burnout in Millennial Medical Students during the COVID-19 Pandemic!" *Irish Journal of Medical Science (1971 -)* 192 (2): 513–519. https://doi.org/10.1007/s11845-022-03016-8

Le Fanu, James. 2002. *The Rise and Fall of Modern Medicine.* 1st Carroll & Graf trade pbk. ed. New York: Carroll & Graf Publishers.

Moses, Jacob D, Agnes Arnold-Forster, and Samuel V Schotland. 2023. "Introduction: Healthcare Practitioners' Emotions and the Politics of Well-Being in Twentieth Century Anglo-America." *Journal of the History of Medicine and Allied Sciences*, May, jrad023. https://doi.org/10.1093/jhmas/jrad023

Nestel, Debra, and Tanya Tierney. 2007. "Role-Play for Medical Students Learning about Communication: Guidelines for Maximising Benefits." *BMC Medical Education* 7 (1): 3. https://doi.org/10.1186/1472-6920-7-3

Nussbaum, Martha Craven. 2001. *Upheavals of Thought: The Intelligence of Emotions.* Cambridge: Cambridge University Press.

Ong, Chong Yau, Mon Mon Yee, Khine Myat Soe, Ruth Elizabeth Abraham, Osca J.H. Tan, and Edmund L.C. Ong. 2022. "Role-Playing in Medical Education: An Experience from Public Role-Players." *Educación Médica* 23 (6): 100767. https://doi.org/10.1016/j.edumed.2022.100767

Parsons, Talcott. 1951. *The Social System.* Abingdon: Taylor & Francis.

Prendergast, Mabel, Alexandra M. Cardoso Pinto, Christopher-James Harvey, and Elizabeth Muir. 2024. "Burnout in Early Year Medical Students: Experiences, Drivers and the Perceived Value of a Reflection-Based Intervention." *BMC Medical Education* 24 (1): 7. https://doi.org/10.1186/s12909-023-04948-0

Reznik, Florence. 2009. "Le groupe Balint, une autre façon de penser le soin." *Le Journal des psychologues* n° 270 (7): 29–30.

Schwartz, Rachel, Tait D. Shanafelt, Christophe Gimmler, and Lars Osterberg. 2020. "Developing Institutional Infrastructure for Physician Wellness: Qualitative Insights from VA Physicians." *BMC Health Services Research* 20 (1): 7. https://doi.org/10.1186/s12913-019-4783-9

Singh, Kirti, Mainak Bhattacharyya, Vikas Veerwal, and Arshi Singh. 2017. "Using Role-Plays as an Empathy Education Tool for Ophthalmology Postgraduate." *International Journal of Applied and Basic Medical Research* 7 (5): 62. https://doi.org/10.4103/ijabmr.IJABMR_145_17

Smith, Allen C., and Sherryl Kleinman. 1989. "Managing Emotions in Medical School: Students' Contacts with the Living and the Dead." *Social Psychology Quarterly* 52 (1): 56. https://doi.org/10.2307/2786904

Stijntje, Willemijn Dijk, Edwin Johan Duijzer, and Wienold, Matthias. 2020. "Role of Active Patient Involvement in Undergraduate Medical Education: A Systematic Review." *BMJ Open* 10 (7): e037217. https://doi.org/10.1136/bmjopen-2020-037217

Taylor, Cath, Andreas Xyrichis, Mary C Leamy, Ellie Reynolds, and Jill Maben. 2018. "Can Schwartz Center Rounds Support Healthcare Staff with Emotional Challenges at Work, and How Do They Compare with Other Interventions Aimed at Providing Similar Support? A Systematic Review and Scoping Reviews." *BMJ Open* 8 (10): e024254. https://doi.org/10.1136/bmjopen-2018-024254

Torppa, Martina A., Eeva Makkonen, Camilla Mårtenson, and Kaisu H. Pitkälä. 2008. "A Qualitative Analysis of Student Balint Groups in Medical Education: Contexts and Triggers of Case Presentations and Discussion Themes." *Patient Education and Counseling* 72 (1): 5–11. https://doi.org/10.1016/j.pec.2008.01.012

Towle, Angela, Lesley Bainbridge, William Godolphin, Arlene Katz, Cathy Kline, Beth Lown, Ioana Madularu, Patricia Solomon, and Jill Thistlethwaite. 2010. "Active Patient Involvement

in the Education of Health Professionals: Active Patient Involvement in Education." *Medical Education* 44 (1): 64–74. https://doi.org/10.1111/j.1365-2923.2009.03530.x
Wadhwa, Anupma, and Sanjay Mahant. 2022. "Humility in Medical Practice: A Qualitative Study of Peer-Nominated Excellent Clinicians." *BMC Medical Education* 22 (1): 88. https://doi.org/10.1186/s12909-022-03146-8

Part III
Conclusion

Chapter 9
Conclusion: Affective Injustice, Empirical Research, and Paths Forward

Abstract In the conclusion to this book, I will discuss the main points and perspectives to come out of the discussion on affective injustice in healthcare. Firstly, I will discuss the necessary conceptual precision needed in this topic, in particular the need to specify affective injustice in relation to the epistemic injustice literature and how we can go about this. Next, I will discuss several recurring themes I have explored throughout this enquiry, namely the equilibrium to be found in emotional regulation strategies and affective expression, and how we can develop greater affective humility in healthcare. I will then advocate that any applications to the illness experience take into account not only the individual but also the collective dimensions, including its impact on care, to understand the extent of the problem. I will also argue for greater attention to marginalized groups and the affective harms they suffer, a group that includes patients but also healthcare providers. Finally, I will discuss the need to move this debate forward with applications to issues specific to certain healthcare providers.

Keywords Affects · Affective injustice · Emotional regulation · Empirical bioethics · Marginalized groups

9.1 Introduction

In this conclusion, I will return to the debates raised in this book and in particular how the preliminary considerations and analyses can be furthered through empirical work. The purpose of this chapter is to provide a way forward both from a theoretical and empirical perspective. Table 9.1 outlines the main subjects that will be discussed.

As suggested in the table, the first issue is the need to take the affective and epistemic injustice framework forward through empirical research. In development thus far of epistemic injustice, the subject has been mainly been elaborated from a conceptual perspective. As I stated in the introduction to Part II, as a non-ideal theory, both affective and epistemic injustice are situated in the real world and its problems. This suggests that in order for this debate to move forward and to find

B. Bogaert, *Affective Injustice in Healthcare*, The International Library of Bioethics 112, https://doi.org/10.1007/978-3-031-94378-2_9

Table 9.1 Ideas to move the affective injustice debate forward

Concept/need	Suggested empirical work
1. Differences with epistemic injustice	Empirical bioethics approaches (normative/descriptive) to explore affective harms from the user perspective and to develop differences with epistemic justice
2. Equilibrium to be found between emotional regulation strategies and the need for emotional expression	Need to determine to what extent healthcare actors suffer affective harms through emotional regulation strategies and in what circumstances they may be beneficial; research to articulate how other persons (patients, families, etc.) see and understand emotional regulation strategies
3. Need to cultivate affective humility in medical students and in healthcare professionals	Initiatives in healthcare institutions and pedagogies in medical training, explored through empirical bioethics research and evaluated through intervention research
4. Better understanding of affective expression and recognition for marginalized groups in healthcare	Approaches to determine how specific patients and healthcare professionals are affectively silenced or misunderstood in function of their social status
5. Nuancing the debate by better understanding the specific issues facing other healthcare professionals	Need to understand the issues facing specific healthcare professionals, such as nurses or nurse assistants, to untangle the link between their social status in the team and the perception of their affects

solutions, we will need to better understand those most affected by epistemic and affective injustice. For instance, there is a need to determine—from healthcare actors' perspectives—how to find the equilibrium between emotional regulation strategies and safe spaces for emotional expression. The best way forward is to understand user perspectives, rather than imposing it from above.

In this chapter, I will principally focus on the benefits of qualitative methodologies to move forward the debate on affective injustice in healthcare. They will be ideal to delve into the micro experiences in which affective harms occur, either by proceeding with case studies or with interviews/focus groups to help participants discuss in what circumstances and due to what reasons it happens. However, to better understand the extent of the problem, quantitative research is also needed to understand how many patients, families, and healthcare providers may be subject to affective injustice in healthcare, both to have a better idea of the problem and to advocate for possible remedial strategies. It is also not to be forgotten, as discussed in Chapters 6 and 8, that the involvement of patients may be a facilitating factor to help some express how they have been harmed in a more open manner. While I have not specifically explored the value of patient research partners in this book, their value in improving research methods and analysis, and in helping research participants to feel comfortable in the interview process, may help overcome silencing and self-silencing in this controversial subject.

9.2 Demarcating Differences Between Affective and Epistemic Injustice

I have advocated throughout this book that affects are a form of knowledge and because of this, affective harms often have epistemic implications. This has been important to my discussion in order to turn the tables on how affects are seen and understood in the context of healthcare and to persuade the reader that they must be taken seriously in medical reasoning, in shared decision making, and in patient appropriations and understanding of illness experiences. However, that does not mean that affective and epistemic injustice are necessarily intertwined, or that affective injustice is merely a side topic of epistemic injustice. As other theorists have highlighted (Gallegos 2021; Archer and Mills 2019), what is involved in affective injustice may include but also go beyond epistemic concerns, in particular as it assesses harms to well-being. As seen in Gallegos' representation of affective injustice (2021), one of the key means to achieve this form of social justice is to provide affective resources for the person's well-being. Many of the proposals of Part II have been focused on this ambition, including through narrative methods, hospital architecture and design, and affective humility. Thus, in my characterization of affective injustice in healthcare, I have also relied on the idea that it involves other harms for healthcare workers and patients. However, there will be a need going forward to delve further into this issue and work out the ways in which affective injustice is different to or complementary from the epistemic injustice debate in healthcare.

I believe that empirical bioethics research can be helpful to better conceptualize and clarify these differences, particularly in healthcare, where the patient, healthcare provider, and family perspective is necessary to better understand these debates. While there is no consensus at the current time, empirical bioethicists currently practice both descriptive and normative approaches (Benetreau et al. 2023). Normative empirical bioethics research has been in particular advocated by theorists such as John Ives and Heather Draper (2009). They have advocated that an interventionist approach is necessary, in which the goal will be to delve into the normative assumptions of the person and a reflective equilibrium reached where the demands of theory and data are weighed carefully, and a compromise reached. This has been opposed to the more descriptive approaches practiced in the social and human sciences, in which the end goal is to describe a phenomenon without intervening directly in the research process or necessarily promoting normative conclusions at the end. In the context of our subject, however, I advocate that both descriptive and normative approaches may be insightful (Bogaert and Dubar 2023) and that when normative approaches are advanced, that they take a form of normative modesty. Spranzi (2018) describes this humbler posture as having a detailed knowledge of contexts of action, contributing to nuancing normative responses in their scope and claim to truth.

Taking into account these possibilities, here are a few suggestions of what empirical bioethics research on affective injustice might look like. One of the ways to go further would be to invest in research from the actor's perspective of what harms they may suffer in having to silence or self-silence their affects. Let us take the example of the nurse who is affectively silenced in an interdisciplinary meeting or who decides in advance to self-silence their affects as they know that they will not be understood or taken seriously. From a theoretical perspective, we know that such situations may cause harms to the person. However, what harms are these? Are they harms to the person's well-being, to their confidence, to their feelings of self-worth? Does it put to task their status and place as knowers? Or are there other harms that we may not be aware of? Empirical research can help us answer these questions.

The researcher may decide to use Ives' interventionist approach (Ives and Draper 2009) by seeking to delve in greater depth into the source of this silencing, to challenge the reasons and assumptions that are behind their use and choice of silence. This kind of approach may help bring to the surface the kinds of implicit assumptions that are behind emotional expression, in particular for those who are heavily socialized at the beginning of medical training. They may also be explored with other members of the team to better understand the assumptions and expectations of how such meetings take place, their ideas on professionalism, and the place for affective expression. This kind of approach may help to identify a number of distinctions and convergences between the theories of affective and epistemic injustice, which may also lead to a number of normative approaches of what to do about it.

However, this topic may also be taken further through a more descriptive approach complementary to social and human sciences approaches. They may involve for instance observation of interdisciplinary meetings to understand in which ways and means affective injustice happens. While directly intervening and delving to the source of the problem has its advantages, direct observation and description can help better understand what happens in practice, the interactions among staff members, and how persons react in function of the social status and individual personalities of each person in the room. Conversational analysis of what exactly happened can also show group dynamics and how conversations and interactions led to affective injustice. In turn, following up these observations with interviews can help engage participants in an active analysis of the researcher's interpretations, including in their own self-identification of the harms and benefits such silences have caused. Another advantage of this kind of research is also that it can easily be done in an interdisciplinary context, for instance in duo with sociologists or anthropologists—which will also help bring forth other complementary or competing theories to move forward this debate, distinguishing it not only from the epistemic injustice discussion but also competing theories and knowledge from the social and human sciences. Although affective justice theorists have sought a reproachment with psychological theories, they have not yet sufficiently widened the debate to other disciplines to see what they can offer to the affective injustice discussion.

9.3 Emotional Regulation Strategies

In this second debate, I will consider the necessary equilibrium to be found between emotional regulation and the need for emotional expression. Throughout this book, I have tried to nuance the debate on emotional regulation strategies of healthcare providers by showing that they may cause important affective harms for them. I have emphasized this point given the extensive literature—in particular in psychology—on the positive influences of emotional regulation strategies for healthcare providers. I have argued that the negative effects of these strategies have thus far been insufficiently considered and may benefit from an analysis under the affective injustice framework. As I have shown in Chapter 2 (on healthcare providers), in Chapter 3 (on patients), and in Chapter 8 (on medical students), there remain a number of troubling emotional regulations strategies that may be harmful for healthcare actors and care provision more generally, in particular those which result in affective silencing and emotional suppression. They rarely deal with the problem at hand and often neglect the person's needs in difficult care situations. As discussed throughout this book, there is a need for greater affective resources and opportunities for these healthcare actors.

From a research perspective, however, there is also a need to better understand how emotional regulation strategies may be perceived by others. For instance, while our discussion has helped understand that it is difficult for healthcare providers to express affects in front of their patients due to their roles as epistemic authorities, little is known on how patients perceive the affects of their doctors. Some preliminary research has suggested that patients consider their doctor emotionless "empty vessels," in that they value less the person behind the doctor, but rather the understanding and recognition of their own emotions (Schroeder and Fishbach 2015). This finding should be further studied with other groups and populations to understand if this finding is specific or applies to wider situations, as it may suggest that emotional regulation strategies in the consultation are valued from the patient perspective. However, even if confirmed, this suggests that patients themselves may be perpetuators of affective injustice toward healthcare providers, including in expectations by them that doctors should suppress their emotions. Thus, there is a need to understand how persons perceive, react, and adapt to emotional regulation strategies in others, including what effects this may have on their relationships and more globally on care delivery.

This kind of research will help to understand in what means, temporalities, and spaces emotional regulation strategies are necessary and what are its positive and negative impact on care. It will also encourage us to find other ways to ensure that healthcare providers may exercise affective resources, opportunities, freedoms, and recognition. While emotional regulation strategies can and will continue to be helpful in care, healthcare providers will need other outlets to deal with, discuss, and take seriously what they have lived, including negative affects like anger and frustration that, should they remain hidden or suppressed, may cause harm to their well-being.

This book has shown the need to provide such institutional support and training mechanisms at the hospital for healthcare providers.

However, as I have advocated, a variety of methods will be needed. One healthcare provider may find benefit from a hospital garden, another from reading and discussing a literary text, another from having an outside resource person with whom they can openly discuss. If we care about healthcare providers' needs, we cannot be content with only one method or way of seeing. In addition, for these resources and opportunities to be fully recognized and valued, they will need to become both part of the permanent hospital landscape, but also valued at the institutional level. This includes enabling healthcare providers to access these resources and opportunities as part of their work time, but also a desire to integrate the learning from them into institutional thinking and change.

9.4 Affective Humility

The concept of affective humility has been important to understanding affective injustice in healthcare. It is both a virtue to be practiced and learned, but also a means to influence how healthcare is perceived and delivered. As discussed in previous chapters, a posture toward the patient that includes an openness to their experiences is one that is centered on the patient. It means accepting that affects are sources of knowledge and that the other person in the relationship will never be able to fully comprehend or share these experiences. While the doctor may have a certain epistemic authority and medical knowledge important for treatment, the patient's experiential knowledge is also to be taken seriously. A disposition to listen to and try to understand the person's experiences in the consultation—to provide that necessary space—will work toward affective, but also epistemic justice. The same process goes for understanding colleagues' experiences.

In this chapter, I have explored some possibilities for cultivating affective humility in medical school training in particular, but I believe that it is an important subject that needs to move into the clinic to be realistic. Here again, empirical research—research from the perspective of healthcare providers and patients—can both help to specify what affective and epistemic humility may look like, but also to move it beyond an ideal. One of the ways to do so will be to focus on patient perspectives and their perceptions on how their affects are solicited, received and understood, as well as to gather healthcare providers' perspectives. We will also need to practice humility ourselves as researchers and avoid presuming in advance that we know in what temporalities and situations patients wish to have their affects known. Research can also help to understand other healthcare providers' perceptions of and attitudes toward humility to clarify how it affects workplace relationships. While this virtue can already be cultivated from medical training, better understanding their perspectives will help us to better understand both how humility may be cultivated—but also practiced in the field. In Chapter 8, I have proposed some promising ways to cultivate affective humility in future doctors. One of these was patient teachers, in particular when a wide variety of perspectives and patient profiles are made available to students.

At the current time, the notion of humility in general (and in medicine in particular) is lacking conceptual precision to move it forward. This means that it has a hard time being integrating as a core medical competency. Conceptual research is needed to make the term more precise, in particular from the perspective of its users—those whom we are asking to practice it. By developing user-centered perspectives of the concept, we will also be better placed to suggest strategies and programs that respond to real needs and are not detached from clinical realities.

9.5 Affective Injustice and Its Individual and Collective Dimensions

Thus far in the epistemic injustice debate in healthcare, characterizations have largely focused on the individual level of analysis, such as doctor-patient relationships. However, scholars are increasingly looking at the institutional and collective dimensions of this phenomenon and ways to combat it, such as the impact of patient access to clinical notes (Blease et al. 2022), calls for greater patient involvement in knowledge creation (Bogaert 2020) or hierarchical issues in healthcare teams (Reed and Rishel 2015). The debate on affective injustice in healthcare will also need to take on both this individual and collective dimension. As Fricker (2009) showed in her original characterization of epistemic injustice, while effects such as silencing will likely have a negative effect on the individual person concerned, they also can influence entire epistemic communities as well as have a negative effect on overall society. Stockdale (2024) has also argued for an analysis of affective injustice as a structural phenomenon, given that it operates primarily in norms, practices and relationships that are embedded in social institutions. As I have shown in this book, while affective injustice often happens in interpersonal relationships, it is also an institutional problem, as it affects quality of care. It is also a structural one, as it is part of the exclusion facing certain members of society in particular. This means that affective injustice must take an interpersonal, individual approach but also one turned toward a more structural perspective.

In scholarship going forward, this individual and collective dimension will need to be analyzed. For instance, if we want to explore if cancer patients are subject to affective injustice, it will be necessary understand the complexities of how they are received by individual care providers (whether emotional work is delegated, if their concerns are taken seriously and understood), but also how such dimensions may affect their perceptions of care, satisfaction, and their health outcomes. In a similar way, as the experience of illness is not confined to the hospital, a better understanding of how their affects are received by their family members, colleagues, and friends will help us to better understand the extent of affective injustice and how it influences the overall illness experience. Here again, empirical work—and in particular qualitative methods—are best suited to this endeavor. They will permit persons to talk about the experiences most salient to them and put into words the affective harms caused, as well as help better understand what strategies they undertook to deal with them. In addition, widening these strategies to a wider group of patients (through quantitative methods) will help understand the extent of the problem for a certain group of patients. Finally, making the link between affective injustice and care quality will help policy makers and hospital decision-makers take the question seriously, rather than relegating it to a few patients' subjective experiences in healthcare.

9.6 Affective Injustice and Marginalized Groups

At the heart of the epistemic and affective injustice discussion has been an ethical commitment turned toward those most excluded in society. In Chapter 3, I have paid particular attention to the affective injustices lived by certain groups due to prejudice, while in Chapter 5, I discussed how spiritual care may be more responsive to this group's affective needs. However, as this book was a first attempt to understand the ways and means in which healthcare providers, patients, and families may suffer from affective injustice in healthcare, it took a more general stance and did not look in detail at the specific issues affecting marginalized groups in particular, apart from a few examples. Going forward, it will important both to understand the extent of affective injustice that certain persons experience in healthcare due to wider social factors of prejudice and exclusion, and to understand how it leads to inadequate care.

When exploring such issues in the future, however, we must not limit this analysis to patient experience (alone). The majority of the epistemic injustice literature applied to healthcare thus far has focused on the patient's experience, and while this is an important as a means of understanding the difficulties of having the patient being heard and understood in medicine and how it is practiced today, most have not focused on healthcare provider experiences and the difficulties they face in being heard and understood either by patients or by colleagues and institutions. The growing literature on the discrimination faced by healthcare providers (Dyrbye et al. 2022) has shown that they too are subject to exclusion based upon their social status as a marginalized group. This includes being unfairly judged by patients, healthcare providers, and institutions. It will be worth exploring how epistemic and affective injustice happens in practice to healthcare providers from minority groups, and how this affects healthcare quality overall, staff retention, and burnout. While this book was unable to treat these issues, they merit exploration both from an affective and epistemic injustice angle, including the identification of possible solutions.

Finally, in Chapter 2 of this book, I could only dedicate a small section to the specific affective harms experienced by paramedical professionals, and in particular the ways I believed nurses suffer from specific affective harms. I hope scholars will work on these issues in the coming years to have both a more holistic understanding of affective injustice in healthcare work—but also to understand the specific stressors and motivations that each professional may face. It will also be worth exploring the intersectional issues of affective injustice in terms of the care role and other identities. It is in this way that we can take care of all healthcare professionals.

9.7 Conclusion

In this concluding chapter, I have given some ways of taking forward the affective injustice debate both from a theoretical and empirical perspective. While I have focused primarily on healthcare, future explorations—in particular empirical research—will undoubtably help move forward both the overall affective injustice debate, but also likely the epistemic injustice one as well. Indeed, the greatest strength of the epistemic—and now I hope affective injustice—debate is that it calls for real life solutions and the involvement of multiple actors to rectify the problems observed and shown. It also reminds us that we are all—whether healthcare workers, researchers, patients, or citizens—responsible to ensure that epistemic and affective justice is made possible for all.

References

Archer, Alfred, and Georgina Mills. 2019. "Anger, Affective Injustice, and Emotion Regulation." *Philosophical Topics* 47 (2): 75–94.

Benetreau, Maud, Marion Bérard, Brenda Bogaert, Margaux Dubar, and Damien Delorme. 2023. *Manifeste pour une philosophe de terrain*. Philosophie de terrain. Dijon: Presses Universitaires de Dijon.

Blease, Charlotte, Liz Salmi, Hanife Rexhepi, Maria Hägglund, and Catherine M DesRoches. 2022. "Patients, Clinicians and Open Notes: Information Blocking as a Case of Epistemic Injustice." *Journal of Medical Ethics* 48 (10): 785–793. https://doi.org/10.1136/medethics-2021-107275

Bogaert, Brenda. 2020. "Need for Patient-Developed Concepts of Empowerment to Rectify Epistemic Injustice and Advance Person-Centred Care." *Journal of Medical Ethics*, November, medethics-2020–106558. https://doi.org/10.1136/medethics-2020-106558

Bogaert, Brenda, and Margaux Dubar. 2023. "Philosophie de terrain en France et à l'étranger : une proposition de dialogue méthodologique." In *Le soin en démesure: alternatives pour des approches qualitatives de la relation de soin*. Paris: Éditions Seli Arslan.

Dyrbye, Liselotte N., Colin P. West, Christine A. Sinsky, Mickey Trockel, Michael Tutty, Daniel Satele, Lindsey Carlasare, and Tait Shanafelt. 2022. "Physicians' Experiences With Mistreatment and Discrimination by Patients, Families, and Visitors and Association With Burnout." *JAMA Network Open* 5 (5): e2213080. https://doi.org/10.1001/jamanetworkopen.2022.13080

Fricker, Miranda. 2009. *Epistemic Injustice: Power and the Ethics of Knowing*. 1. publ. in paperback. Oxford: Oxford University Press.

Gallegos, Francisco. 2021. "Affective Injustice and Fundamental Affective Goods." *Journal of Social Philosophy* n/a (n/a). https://doi.org/10.1111/josp.12428

Ives, Jonathan, and Heather Draper. 2009. "Appropriate Methodologies for Empirical Bioethics: It's All Relative." *Bioethics* 23 (4): 249–258. https://doi.org/10.1111/j.1467-8519.2009.01715.x

Reed, Pamela G., and Cindy J. Rishel. 2015. "Epistemic Injustice and Nurse Moral Distress: Perspective for Policy Development." *Nursing Science Quarterly* 28 (3): 241–244. https://doi.org/10.1177/0894318415585634

Schroeder, Juliana, and Ayelet Fishbach. 2015. “The ‘Empty Vessel’ Physician: Physicians’ Instrumentality Makes Them Seem Personally Empty.” *Social Psychological and Personality Science* 6 (8): 940–949. https://doi.org/10.1177/1948550615597976

Spranzi, Marta. 2018. *Le Travail de l’éthique: Décision Clinique et Intuitions Morales*. Psy—Théories, Débats, Synthèse. Sprimont: Pierre Mardaga Editeur.

Stockdale, Katie. 2024. “(Why) Do We Need a Theory of Affective Injustice?” *Philosophical Topics*. 51 1.

MIX
Papier aus verantwortungsvollen Quellen
Paper from responsible sources
FSC® C105338

If you have any concerns about our products,
you can contact us on
ProductSafety@springernature.com

In case Publisher is established outside the EU,
the EU authorized representative is:
Springer Nature Customer Service Center GmbH
Europaplatz 3, 69115 Heidelberg, Germany

Printed by Libri Plureos GmbH
in Hamburg, Germany